The Lost Treasure of King Juba

The Lost Treasure of King Juba

The Evidence of Africans in America before Columbus

FRANK JOSEPH

Bear & Company
Rochester, Vermont

Bear & Company
One Park Street
Rochester, Vermont 05767
www.InnerTraditions.com

Copyright © 2003 by Frank Joseph

All rights reserved. No part of this book may be reproduced or utilized in any form or by any means, electronic or mechanical, including photocopying, recording, or by any information storage and retrieval system, without permission in writing from the publisher.

LIBRARY OF CONGRESS CATALOGING-IN-PUBLICATION DATA

Joseph, Frank.
  The lost treasure of King Juba : the evidence of Africans in America before Columbus / Frank Joseph.
    p. cm.
  Includes bibliographical references.
  ISBN 1-59143-006-2 (pbk.)
  1. Africans—Illinois—Antiquities. 2. Blacks—Illinois—Antiquities. 3. Treasure-trove—Illinois. 4. Caves—Illinois. 5. Illinois—Antiquities. 6. America—Discovery and exploration—African. 7. Mauretania (Kingdom)—History. 8. Mauretania (Kingdom)—Antiquities. I. Title.

F550.A24J67 2003
970.01'9—dc21
               2003000130

Printed and bound in the United States at Lake Book Manufacturing, Inc.

10   9   8   7   6   5   4   3   2   1

Text design by Cynthia Ryan Coad
Text layout by Virginia Scott Bowman
This book was typeset in Bembo and Gill Sans with Packard and Agenda as the display typefaces.

All photos courtesy of *Ancient American* magazine except where captions indicate otherwise.

*To Wayne May, without whom
nothing would have happened*

Contents

	Preface	ix
	Introduction: A Shattering Revelation	1
1	It All Started with Cleopatra	3
2	Mauretania	21
3	SPQR: For the Senate and People of Rome	39
4	Caligula: A Mind Abused	46
5	Claudius: The Failed Peacemaker	52
6	Escape or Die	59
7	Discovery in Southern Illinois	69
8	Gold, Archaeological and Otherwise	78
9	Find or Fraud of the Century?	90

10	Fire in the Hole	115
11	Where Is the Cave?	136
12	The Pasture of Fools	150
13	The Testimony of the Past	165
14	A Rooster Speaks	174
15	Lost Coins and Buried Treasure	199
16	Epilogue: The Moment of Truth	215
	Appendix 1: A Mauretanian Time Line	219
	Appendix 2: Ancient Stone Maps	223
	Notes	229
	Bibliography	237

Preface

As the editor in chief of *Ancient American* magazine, I am regularly prevailed upon by mostly amateur archaeologists to publish their reports of overseas visitors to our continent during pre-columbian times. While many of these submissions may be interesting, sometimes provocative, they are usually unsubstantiated by any credible physical evidence. A story first brought to my attention in 1993, however, was supported by an abundance of material items—more than seven thousand, in fact. The sheer volume of such alleged proof combined with the often superb workmanship of numerous individual pieces argued persuasively on behalf of their authenticity.

Even so, I was baffled by not only the magnitude of the discovery, but also the profusion of its disparate cultural imagery. How was one to account for images that appeared to be Romans, Celts, Christians, Jews, West African blacks, Egyptians, and Phoenicians all represented together at a single, subterranean site in, of all places, southern Illinois? Over the next nine years, I not only described the Burrows Cave collection in several feature *Ancient American* articles, but also undertook my own investigation of the supposed artifacts to determine their authenticity, at least to my own satisfaction.

The conclusions of various authorities in mineralogy and ancient written languages I consulted suggested that retrieval of the seven thousand images found near a tributary of the Ohio River represented the greatest archaeological event in history, far more spectacular than the opening of Pharaoh Tutankhamen's tomb sixty years before. The Illinois tomb not only appeared to contain vaster amounts of buried treasure, but, more valuably, also demonstrated that Roman-era visitors crossed the Atlantic Ocean to establish a settlement in North America nearly fifteen centuries before Columbus sailed from Spain. Although the saga of these voyagers is far from completely understood and the unveiling of Burrows Cave signifies a work in progress, both are aspects of a story that must be told.

I have combined years of my own research with the expertise of professional scientists and enlightened enthusiasts alike to create a mosaic from different fragments of evidence. Bringing these pieces together—fitting them into a complex archaeological puzzle—was my purpose in writing this book. Through its pages march heroes and villains, tyrants and freedom fighters, mystics and profiteers, victors and survivors. Their story is valuable because it is our story, lost for the last two thousand years but now gradually coming to light from its underground burial sanctuary. With its retelling, the dead will live again, and the roots of American history, far deeper and older than suspected, stand revealed.

Introduction

A Shattering Revelation

*It is the dead who have a tale to tell—the dead who
died centuries ago—to people who still live.*
COUNT BYRON DE PROROK,
IN QUEST OF LOST WORLDS

There are some disclosures that radically revolutionize long established conceptions of the world in which we live. The subject of this story is one of them, because it demolishes what Americans have been led to believe since their country was founded—namely, that Christopher Columbus was its discoverer. An archaeological cave site in southern Illinois suggests instead that tens of thousands of refugees sailing from the murder of their king and the invasion of their homeland preceded Columbus by nearly fifteen centuries. Preferring a perilous transatlantic adventure to slaughter and slavery on land, they entrusted their lives to the sea.

There is a contemporary side to this tale. It tells of the cave's discovery, a subsequent twenty-year period of imposed secrecy, the looting of the cave's fabulous treasures, an often bitter controversy, and final disclosure. But the first part of this story is much older. It describes what was formerly a splendid kingdom in the ancient world, a vital part of the Roman Empire that was once culturally rich and economically powerful, but which was reduced to obscurity by war. Faced with the choice between almost certain death at home and escape over the uncertain open sea, some of its survivors became first-century "boat people." Most successfully completed the crossing to America only a few years after the death of Jesus.

Although the majority of professional archaeologists dismiss such transatlantic voyages as imaginative fantasy, they are contradicted by a vast collection of inscribed and illustrated stone tablets uncovered from a subterranean site in the American Midwest. Often wonderful masterpieces of art, they comprise thousands of portraits of men and women from a distant land in ancient times. There are grim-faced soldiers and sagacious priests, sailors and worshipers, kings and queens. They are accompanied by tablets inscribed in several different languages, some of which have already been partially translated. And there is gold, a treasure trove King Solomon in all his splendor would have envied.

Both stories seem too fantastic for belief. Yet, an abundance of hard and historical evidence supports their credibility. The fabulously rich legacy buried nearly two thousand years ago was known only to the elders of a particular Indian tribe, whose last chief broke the secret before he passed away. Even then, the whereabouts of the cave were unknown until it was found by accident twenty-four years later. The sometimes acrimonious struggle to open the site and unravel its significance has lasted almost as long. That struggle still goes on. But the time has come for its story to be told.

1
It All Started with Cleopatra

*Th'abuse of greatness is when it disjoins
Remorse from power.*
SHAKESPEARE, JULIUS CAESAR 2.1

Most of the artifacts removed from a subterranean location in southern Illinois since 1982 comprise the portraits of Romans, black Africans, Egyptians, Phoenicians, Jews, early Christians and Native American Indians—all, judging by their attire, from ancient times. Almost invariably, they were accompanied by written languages in hieroglyphic Egyptian, North Semitic (or Carthaginian), paleo-Hebrew, ancient Iberian, Ogham, or an unknown script found nowhere else. Most of the faces are depicted in profile, and the majority of these belong to Roman-style soldiers. Other perceived professions include holy men and sailors. Far fewer women and elderly persons appear, and no children are represented.

Religious imagery includes the so-called Alexander Helios symbol, the circle cross, Jewish menorahs, Stars of David, Christ-like figures, Egyptian pyramids, the Greek Pan, the Carthaginian Tanit, and other, less identifiably pagan creatures. Other tablets are given over entirely to lengthy, largely untranslatable inscriptions or the depiction of ancient sailing ships. Some are Phoenician, but others resemble Roman vessels. Animals frequently portrayed are cows, rams, elephants, serpents, whales, fish, sea monsters, and other fabulous creatures, sometimes half human.

For many years after these strange illustrated objects were brought to the attention of scholars, even those who granted the possibility of overseas visitors to the Americas during pre-columbian times threw up their hands in disbelief at the impossible variety of disparate races and religions represented at the same site. Nothing seemed able to explain such an incongruous jumble of unrelated cultures and peoples, especially in, of all places, Illinois, many hundreds of miles from the nearest sea coast. The collection had to be fake. But the sheer number of its objects and the frequent excellence of their execution were in sharp contrast to the inconceivable implications of their origins.

A single piece among the estimated seven thousand controversial artifacts, however, is the first of many clues to the origins of the collection. This decisive item is a gold medallion about the size of a dollar coin (although three times as thick), struck with the image of an elephant's head (fig. 1.1).

Fig. 1.1. A gold coin from the Illinois cave featuring the image of an elephant's head, which was the personal emblem of Cleopatra VII, the Ptolemaic queen of Egypt in the first century B.C. Photograph by Beverley Moseley

Identical gold pieces were minted more than two thousand years ago by Cleopatra VII, the Great, when she became the famous queen of Egypt. She chose the elephant head as her personal emblem for cogent political reasons.

Cleopatra was descended from a long line of Ptolemaic rulers established after the death of Alexander the Great, nearly three centuries earlier. The queen belonged to this Greek ruling class, which she dominated from 48 to 30 B.C., and was, therefore, without a drop of Egyptian blood in her veins. Because her subjects sometimes chafed under the domination of these foreigners, in selecting the elephant's head as her own insignia Cleopatra was making a powerful visual statement that she was in spirit, if not heritage, entirely African. In reality, she was much more than that.

The queen's circuitous way to the throne, literally over the dead body of a younger brother, had been achieved via her bed. By that time, the seductive arts had reached high levels of application in Egypt with the proportionate decay of that civilization, and the precocious teenage Cleopatra was adept at exercising their power over a susceptible middle-aged Julius Caesar (fig. 1.2). For two thousand years her name has been synonymous with the height of feminine allure. Beginning in the mid-twentieth century, however, when the deprecation of all things perceived as Western became fashionable, she was portrayed as dowdy. More recently, she has become militant feminism's historical icon, tragically betrayed by loutish, devious men. The truth, as usual, lies somewhere in between these extremes. Contemporary and postmortem statuary portraits of Cleopatra reveal no great beauty, even given their idealization. They nonetheless credibly depict a slender, attractive woman with a pleasing countenance, a somewhat large though well-formed nose, full cheeks, and an intelligent brow (see fig. 1.3).

Fig. 1.2. Julius Caesar as he appeared when he first met Cleopatra (Capitoline Museum, Rome)

Doubtless, Cleopatra's attraction was a combination of her abilities as a witty and cultivated woman as much as those of a sexual athlete. The erotic spell she cast over Caesar had been politically conjured to secure for her the throne of Egypt. Cleopatra was a clever, ambitious manipulator, and sex was just another tactic in her overall strategy to secure sovereign power. While controversy may still surround her physical endowments, the brilliance of her mind has been beyond question. She alone of all the Ptolemies was fluent in many languages, including Egyptian (the tongue of their own subjects, which few of them deigned to learn), Ethiopian, Arabic, Syrian, Parthian, and Medean. Widely

Fig. 1.3. A first-century B.C. statue, considered one of the most accurate, of Cleopatra VII

traveled throughout the Nile Valley and the eastern Mediterranean world, she took her royal education seriously, and was known as a fast learner.

Naturally skilled and professionally trained in the arts of diplomacy and government, Cleopatra was conversant in the leading philosophical arguments of her day. Her knowledge of naval warfare prompted her to build up the Egyptian fleet. Strong-willed, she nevertheless often yielded to compromise, which enhanced her reputation as approachable and fair-minded. For her, though, apparent conciliations were inevitable trade-offs in diplomatic maneuvering toward the fulfillment of her personal agendas, whose final goals were all that really mattered. But what did she really want? She desired more than just the crown of a superannuated Egypt. Like many potentates, from Alexander to Ghengis Khan to Joseph Stalin, Cleopatra was obsessed with the Eastern dream, known in her day as Hellenism. Its proponents envisioned a world dominated by the cult of the Egyptian mother-goddess Eset, or Isis, as she was known to the Greeks. And it was the Greek Cleopatra who, as the reigning monarch of Egypt, was revered, not just as the cult's high priestess, but also as the actual goddess herself in human flesh.

Such blasphemous presumption had nothing in common with the traditional worship of Eset, which preceded the Ptolemies by almost three thousand years. Eset was one of mankind's most endearing spiritual conceptions, the divine personification of wifely devotion and mother love, with emphasis on the human soul's evolution through compassion and devotion. It was the Ptolemies who used this mystery religion to cover a political movement aimed at the acquisition of earthly domination. Citing the German historian Otto Kornemann, American author Beatrice Chanler writes that he "attributes to the Ptolemies the ambition to extend their power to the farthest limits of the inhabited world. Like Alexander, they dreamed of universal empire."[1]

Cleopatra's millions of hysterical followers prayed for the day when she, the living Isis, would gather up the masses of the East and hurl them against the hated Romans to create a single authority, with herself as the Queen of Heaven and Earth. She had inherited and modified Hellenism, along with her dynasty, from Alexander III (another "Great"), who tried to conquer the world and dominate it under a single system fashioned after Eastern absolutism. To most of his followers, he betrayed his Spartan upbringing—an Aristotelian education emphasizing individual liberty and self-responsibility—by backsliding into Oriental autocracy. After all, his conquests had been initially propelled by the Greek will to civilize the rest of the world. That was the original meaning of Hellenism, and it was what he and his warriors originally fought for. Following years of triumph on behalf of their idea of enlightment, however, he was transformed into its polar opposite, styling himself a living god, demanding that his officers and men prostrate themselves in the dust before him in the manner of all Eastern despots, even insisting they take Persian wives.

"His original idea," Chanler explains, "was to efface in a universal empire the difference between peoples and to melt into a unity of one common civilization the traditions which had been divergent for centuries."[2] It was the *Heart of Darkness* syndrome, in which the conqueror himself, his better judgment obscured by an inflated ego, is subverted through pity for the very people he came to dominate. To put it bluntly, his rational faculties had been eroded after too many years in the field. Great commanders long after him, such as Napoleon Bonaparte and Erwin Rommel, suffered similar psychological deterioration because their nerves were also subjected to the high stress of over-long military campaigning. "Although many since the great general had taken up his idea of empire," writes Chanler, "no one before Cleopatra would take it up with the far-reaching idea which Alexander attached to it: under one absolute authority, of divine right, the entire world must be unified."[3]

Like Alexander the Great, Cleopatra considered herself a living deity. Her name, translating as "glory to her father," was as much a deliberate tribute to Alexander, if not more so, than to her own natural parent. The

Ptolemies were an incestuous lot, and generations of inbreeding resulted in a family megalomania that fueled old notions of world rule. It was not for nothing that the French historian Bouche-Leclercq referred to Cleopatra as a "venomous flower blossoming on an unhealthy stem."[4] Her dynasty built a Soviet-style, centralized government in which agriculture, commerce, and banking were directly controlled by the state. Revenue authorities working for the royal house were armed and aggressive and were given extraordinary powers to collect taxes. Artists of all kinds were government-funded propagandists for the royal house, while favorable investment opportunities were extended to wealthy tradesmen (invariably at the expense of native Egyptians) from all over the Mediterranean and the Middle East. Indeed, in order to encourage foreign loans and bank and land deals, there were no proscriptions against race or religion, making Alexandria the second richest capital; only Rome was wealthier.

In this apparent openness, too, the queen was politically motivated. Persecuted everywhere else, Jews flocked from all over the known world to Egypt, where they were free to practice their beliefs and conduct business. In fact, by Cleopatra's time, no less than a quarter of Alexandria's population was Jewish. This Jewish community had by then become an important part of the burgeoning prosperity of Ptolemaic Egypt. Some Jews rose to become very influential at court, albeit outside the innermost circles, with important financial connections to foreign kingdoms, where their relatives acted as stock agents, particularly in grain trading—all facts not lost on the queen. She learned to speak fluent Hebrew, and among her first acts as queen was to command the construction of a new city synagogue at state expense. And yet, for all this seeming favor, many if not most of Cleopatra's tax police were Jewish, which, not surprisingly, engendered violent anti-Semitism among resident Greeks and native Egyptians for generations thereafter.

Prefiguring Marx's *Communist Manifesto* or Orwell's *1984,* the Ptolemaic motto ran, "No one has the right to do what he wishes, but everything is organized for the best."[5] Most Egyptians made only a subsistence living, however, and were excluded from all positions of political or economic power. Cyclone Covey, professor of history at Wake

Forest University in North Carolina, points out that "Alexandria was notoriously cosmopolitan, with native Egyptians as second-class citizens in their own country. Macedonians and Greek *koine*-speaking Jews comprised the Alexandrine aristocracy . . ."[6] Misery and discontent were common, as were revolts. An underground of patriots dedicated to the Ptolemies' overthrow and the restoration of ancient values was active throughout the Nile Valley. But they had missed their time by a long shot. It was too late to bring back the glory of the pharaohs.

Even so, Alexandria offered numerous public amenities, such as general health care and vigorous trade, with special emphasis on the arts and entertainment. Cleopatra's Egypt was a contrast between fabulous wealth concentrated in the hands of her ruling family (and international merchants) and widespread poverty. To distract the masses from their deplorable conditions and stir patriotic fervor, she often staged magnificent public parades and demonstrations. These were mobile pageants involving thousands of strange animals, brightly costumed actors, singers, and musicians, all performing among spectacular special effects and historical reenactments on long trains of gigantic floats.

These outlandish festivals invariably featured a grand procession for Dionysus, the deified personification of Hellenism, in which a huge map of the world was spread out before a colossal statue of the god. This dramatization always drove the crowds of onlookers to a frenzy of Hellenistic mass hysteria. They recognized that the Dionysus procession exemplified Ptolemy's prophecy, a famous, trumped-up piece of propaganda stating that the dynasty was destined one day to lead Greco-Egypt in unifying the East for the creation of a new order on Earth. Though such public extravaganzas may have provided some emotional outlet or cohesion for a society beset by internal discord, the prophetic fulfillment of world conquest in the late first century B.C. was nowhere in sight. The numerous kingdoms of the eastern Mediterranean had long before shattered into impotent fragments, while Egypt was hardly more than a shadow of her pharaonic past. If anybody was doing the conquering, it was Rome, and she alone.

Ptolemy's prophecy was nothing more than an age-old longing for universal domination through the whims of some deified potentate. In contrast the new Roman concept of individual rights and liberties guaranteed by the state was an innovation of the highest degree in a world otherwise dominated by unrelieved despotism. Eastern autocrats regarded as the most insane subversion the very notion that common people were "citizens" with the power to vote. In Rome freedom and justice were possible only through institutions of law and government upon which Western civilization, even down to the continued use of Latin in our courts, was to persist over the next twenty centuries. America's founding fathers used the Roman model, above all other examples, as the historical template for their constitutional republic, even to the inclusion of the *fasces* (a bundle of rods around an ax—the Roman symbol of authority) as their judicial emblem. The notion of god-ordained kingship (or queenship) so revered by the Ptolemies and their ilk was repellent to the Roman mind, awakened as it was to the possibilities of individual liberty and human rights. As modern citizens of an authoritarian republic, they preferred the rule of law to that of divinity, and manly leadership to divine kingship.

Even so, Cleopatra was shrewd enough to detect an early opportunity for bringing a Hellenistic world into existence, the first such chance since the campaigns of Alexander the Great. Having seduced Julius Caesar, she gave birth to his child. The boy's mere existence would help set in motion the realization of her dream, which progressed an important step farther when she was invited to live in Rome. She urged her lover to make war on Parthia, a powerful empire, occupying most of present-day Iran. Its conquest would bring millions of Parthians and their subject peoples into the Roman world. She would then Hellenize them through her influence over Caesar and position herself as the living Isis to her millions of fanatic followers. The eventual elevation of their son to the throne would crown all her efforts with total success.

But there was a dynamic motivating Caesar at this time as well. Chanler writes:

> One cannot overestimate the importance and power of the Isiac brotherhood [cult of Isis]. It threatened at one time to become the

principle [*sic*] religion of the world, while it aimed at a more intimate control of the state through its closely knit organization. Cleopatra was an instrument through which this aim might be achieved. And Cleopatra reciprocally made use of the cult and her titular role in it to further her own ambition for world power. It was on entering his office of Aedileship that Julius Caesar first became associated with the strength of the secret societies. When he became dictator and suppressed the so-called democratic clubs, saying there was no place in a well-ordered state for an occult government, he refrained from attacking the Isiac fraternities. In Egypt, he had had the opportunity of discussing with the High Pontiff of Isis the cooperation of the Isiac societies, should he ever embark on his worldwide campaign, so even reckoning without Cleopatra's influence, his sanction and protection of the Isiac cult was understandable.[7]

To be sure, Cleopatra did not introduce the Ptolemaic dream to Rome. She was but its latest and most powerful operative. The Isis cult had been subverting Roman society long before she sank her claws into Julius Caesar. In 58 B.C., for example, the consul Gabinius ordered the destruction of every one of its statues that had been restored since the previous proscription, including all Isiac altars, on the authority of the senate. Just four years later, the senate was forced to again demand that all cult images, public and private alike, be abolished. And another four years after this, the consul L. Emilius Paulus ordered the illegally rebuilt Temple of Isis to be torn down. After learning that no worker was willing to raise his hand against it, the senator stormed down to the building, doffed his robes of office, picked up an ax, and began demolishing the temple himself. In another two years, the Ptolemaic Isis was again being worshiped in Rome. To the Romans, the subversive cult had all the tenacity and resilience of an insect infestation.

What could have accounted for the popular appeal of this alien cult? Chanler explains:

> It will be apparent that in the Isiac doctrines of pity for the lowly, immortality of the soul, and an after-life where virtue is rewarded, as

well as its basic precept that all gods were manifestations of one god, etc., that the Isiac cult was not a pagan one, but a direct forerunner of Christianity. Isis was called the Virgin by the Egyptians, and statues of Isis holding her infant son, Horus, were as prevalent then as the Madonna and Child are now. A world torn with war and discord was ripe for the appearance of a Messiah which these prophecies announced.[8]

However, shrewd Romans saw through them to their underlying political intention—namely, the overthrow of the Roman state, and its substitution by a world tyranny veiled in a religion of universal love. When Caesar spoke seriously of campaigning against Parthia, those who understood his ultimate purpose decided on the most drastic course of action. As Chanler points out, "On the eve of his assassination, Caesar was planning to consummate Alexander the Great's *idée maitresse*: World domination! The first step called for the conquest of Parthia."[9] The French historian Jerome Carcopino wrote that just before he was killed, Caesar planned to demand an authorization for polygamy from the senate so he could wed Cleopatra without breaking his marriage and the law. He planned to jointly rule the world with the Egyptian "Queen of Heaven."[10] In Book IV of his *Lives,* Plutarch mentions that polygamy, an affront to Roman morality, was "not prohibited, but customary for the kings of Macedonia," such as Alexander the Great.[11]

Fig. 1.4. First-century B.C. Roman statue of Isis (Vatican Museum)

Julius Caesar was assassinated in 44 B.C., and although his death was a great setback to the Isiac cult, of which Cleopatra was the most important co-conspirator, it did not die out with him. "Even after the Ides of March," Chanler writes, "the religion had the protection of the dead Caesar. The Triumvirs, probably knowing something of the plans or desires of the late Dictator, decreed in 43 B.C. [that] temples to Serapis and Isis [be built], only five years after the Senate had ordered the demolishing of all temples."[12]

In the civil war that followed Caesar's assassination, Cleopatra sailed away without incident, secure in the possession of her prime objective: the Egyptian throne. "At the death of Julius Caesar," Chanler observes, "with the prospects bright for the succession of Caesarion, as the son of Ra, to the dictatorship, the Egyptian cult had seemed about to reach the summit of secular power towards which it was striving, and through it impose a unity of religion upon the entire world."[13] In this, Cleopatra seemed inadvertently abetted by the Roman victors. With their triumph over Caesar's assassins, they divided the empire in three. The eastern provinces were allotted to Marc Antony (fig. 1.5), whose position as eventual emperor seemed almost assured. Although disliked by most of the cultured and privileged classes for his ribaldry and blunt manner, he was beloved as a genuine soldier by the common people and the army, where real power lay.

Antony liked to eat, sing, march, hunt, get drunk, party and joke with his comrades, and sleep in the open on the cold ground under his robes, as his men did while participating in all their common joys and privations. Antony was a skilled field surgeon and personally administered to the wounds and illnesses of his men, even those of soldiers in the lowest ranks. His fellow officers revered him as, in the words of Plutarch, "the most experienced commander living."[14] His generosity to the troops increased proportionately to his ever-greater acquisition of personal power, and in time he eventually began winning support among some senators, who imagined he might serve as a useful cat's-paw for their own agendas.

Fig. 1.5. Marc Antony fathered the child who may have been the first "sun king" of pre-columbian America. (Capitoline Museum, Rome)

But they were left in the dust by Cleopatra. To her, he was her second chance for the realization of Ptolemy's dream. She would transform Marc

Antony into her instrument for greatness, just as she had Caesar. To make certain, she introduced herself in one of the grandest entrances of all time. The queen came sailing up the river Cydnus, in Cilicia, today's southeastern Turkey, on a gigantic barge. Ensconced under a gauzy canopy of golden cloth, she was dressed as Venus, the goddess of love, and was fanned with ostrich feathers wielded by boys painted to resemble cupids. Huge purple sails and silver oars propelled the immense vessel, its decks populated by beautiful attendants dressed as sea nymphs. The entire stern was sheeted in beaten gold, and clouds of lotus incense, together with the music of flutes, fifes, and harps, wafted to the shore, attracting great crowds of astounded onlookers.

Far from any amorous intentions, Antony had come to question Cleopatra about the aid and comfort she allegedly provided Caius Cassius, one of Julius Caesar's assassins. He was prepared to arrest her on serious charges of conspiracy to commit murder, if his suspicions merited such drastic action. Her initial reluctance to answer his written requests for an audience suggests that the charges against her were not without some foundation. Only after repeated assurances by Dellius, the Roman representative, that Marc Antony meant no harm did she set out on her voyage of conquest. Hedging her bets, she doubtless gave tacit support to Cassius in the event that, should he prevail in the civil war that arose from his participation in Caesar's assassination, she might mold him into her love slave. She dropped him as soon as his cause appeared lost.

In any case, allegations of complicity with the murderers were utterly eclipsed by her remarkable appearance. Its effect on Antony was exactly what she hoped to achieve. All his life he had shown many positive, valiant, even superior traits. But from early youth, he also exhibited a weakness for occasionally excessive self-indulgence. Although his willingness to let the good times roll with as many friends as possible somewhat mitigated the worst effects such behavior might have on his reputation, his philandering with married women and drunken debauchery offended even his closest associates. A particularly humiliating instance took place, when, after an all-night drinking party, he vomited in full view of a large crowd while trying to give an important public speech.

Happily, these lower tendencies were beginning to recede as he conscientiously prosecuted the war against Caesar's assassins and executed his governmental responsibilities, which he took very seriously. People began talking of his eventual rise to emperor. His popular following was great and growing. But Cleopatra resuscitated the man's weaknesses, plying him with sex and sumptuous banquets of boar and peacock on gold and silver plate, followed by rounds of raucous parties awash with liquor, all amid scenes of luxury unrivaled anywhere else, even in Rome. She constantly flattered him and commanded all Egypt to do the same, appealing to his male vanity on a royal, even international scale. Plutarch disdainfully reported, "The word went through all the multitude that Venus had come to feast with Bacchus for the common good of Asia." Already, even here, Ptolemy's prophecy was hinted at in Antony's celebration with Cleopatra "for the common good of Asia."[15]

Yet, in the midst of this ongoing bacchanalia, he was shocked back into reality by a dispatch from Rome. His wife, Fulvia, had engineered an armed revolt against his co-regent, Octavian. Her intentions were less political than marital: She hoped that insurrection would cause her husband's break with and return from Cleopatra. He immediately took ship for home, but was informed en route that Fulvia had died of an illness incurred while she was on her way to their reunion. Proceeding in sincere guilt-stricken mourning to Italy, he was met by Octavian, who blamed him for none of his late wife's political indiscretions, urging him instead to take up the fallen reins of good government on behalf of Roman unity, still uncertain after the civil war.

Antony's better nature responded to Octavian, and the two men formed a triumvirate with their comrade Lepidus, who took charge of Africa. Octavian administered the Western world, while Antony controlled the East. To solidify their bond, Antony married his friend's older sister, Octavia. She was in all respects an ideal Roman woman—beautiful, highly cultured, family-oriented, strong, honorable, responsible, generous, and pious. Octavia exercised a gently salubrious effect on Antony, dispelling his old wildness and nurturing his nobility of soul. She genuinely loved him, and he at least deeply honored and respected her. As long as her higher influences surrounded his daily life, Marc Antony's virtues and popularity grew hand in hand, and the world seemed headed toward a *Pax Romana* in which he played a significant role.

When word reached Cleopatra of his marriage to Octavia, she collapsed in grief. All her hopes and plans to become the queen of the world had evaporated. She had not long to wait, however, before the broken thread of her web would be reknit. The Parthians were rattling their sabers again, but Antony, afraid he might yield to temptation if he ventured too near his former lover's sphere of influence, sent Ventidius, his general, to restore order.

Unfortunately, Ventidius was unable to cope with the situation, so Antony went in person, taking with him Octavia and their newborn daughter as safeguards against temptation. While traveling through Greece, they learned that the crisis was more serious than anticipated. Octavia and the child returned to Rome. In Syria, Antony drilled the two legions Octavian sent him for a major military confrontation, while simultaneously dispatching envoys to sound out the Parthians on a diplomatic solution. Memories of Cleopatra, however, began to haunt his every dream and waking hour. They eroded his willpower and good intentions. He knew he should return home to his wife and daughter, but duty prevented him from leaving Asia.

Torn between devotion to the loyal Octavia and lust for the conniving queen, he at last gave way to his baser instincts. An impatient Cleopatra wasted no time, demanding new territories to partially compensate her for the public humiliation she had endured because of his marriage to Octavian's sister. To calm her, he gave her much of Arabia and even a rich slice of Judea. When Antigonus, the Jewish king, strenuously objected to this unwarranted concession, if only because his country was then an ally of Egypt, Antony had him arrested, then, at Cleopatra's urging, beheaded. The senate and people of Rome were astounded by such behavior, but his anxious supporters assured them that, however unorthodox such actions seemed, they were part of everyday Eastern politics.

To forestall the groundswell of hostility beginning to mount against her philandering husband, Antony's courageous wife, on her own initiative, raised two thousand elite troops, fully armed and outfitted into praetorian cohorts, together with baggage, cattle, money, and additional uniforms for his troops. She was personally bringing all of these soldiers and supplies to Antony when she received a letter from him telling her to wait in Athens until he arrived. He was on the point of leaving when Cleopatra feigned a sickness unto death for love of him. He allowed himself to be convinced by her treachery, and dishonorably lingered with the home-wrecking queen.

Octavia returned sadly to Rome but refused to say a word against her distant husband, defying her powerful brother's command that she no longer reside at Antony's house. She continued to lovingly care for his children, even those by Fulvia, and could not be moved to criticize Antony in his absence. But Octavia's quiet sorrow and dignified deportment spoke more eloquently against him than any speech she could have ever delivered. Even his strongest supporters began to melt away, while popular hatred for the man and his "Egyptian whore" simmered, then boiled.

No one did more to exacerbate their rancor than Marc Antony himself. He allowed himself to become the focus of an immense public mass rally at a military exercise ground, where a tall platform of silver was surmounted by a pair of golden thrones. Beneath these, four smaller gilded chairs formed a line. At the dramatic height of the day's elaborate festivities, Antony mounted the platform with Cleopatra, who was dressed to impersonate Isis in a form-fitting, full-length golden gown suggesting feathers, and the couple were seated. He announced that her offspring were being given personal jurisdiction over the East. One by one, the children sat in their little thrones as Antony proclaimed that Caesarion, her son by Julius Caesar, was to rule conjointly with his mother over Egypt and Cyprus. Ptolemy Philadelphus, Cleopatra's youngest child, by Marc Antony, who was given Phoenicia and Cilicia, appeared dressed in boots, mantle, and Macedonian cap adorned with a diadem, the attire of a legitimate successor to Alexander the Great.

Their two other children—twins, a boy and a girl—were particularly special, born moments after dawn on the morning of the winter solstice, December 25, 35 B.C. Cleopatra gave birth to them in the Alexandria palace bedroom suite overlooking the island of Antirrhodus. Just then, a huge eagle, a species never before seen in Egypt, perched on the palace roof. To the priests of Isis from Pharos Island, the ominous bird was an unmistakable sign from the gods that one of the twins would someday preside over a great kingdom. Indeed, the unusual circumstance of their birth was widely regarded as a portent of the highest significance, divine proof that the Ptolemies were destined to rule the world.

In view of the double-solstice birth, the twins were named Cleopatra Selene and Alexander Helios—Cleopatra of the Moon and Alexander of the Sun. Citing German historian Franz Boll, Chanler writes, "These names signify 'Cosmoscratores' (or the two powerful planets, Sun and Moon); and one may consider the choice of these names as the avowal of the Roman Triumvir that he [Antony] destined the terrestrial world for his children by the Queen of Egypt."[16]

Cleopatra VII was now truly Cleopatra the Great. Thanks to her, Ptolemy's prophecy was being fulfilled. She looked serenely toward ruling the planet from Rome itself, where the mystery cult of Isis, of which she was high priestess and the deified embodiment of the goddess, had a large, popular following. At the Alexandrine spectacular in which Marc Antony was so generous with Roman territories and sovereign kingdoms, little Alexander Helios was betrothed to Princess Iotapa, the Median king's five-year-old daughter, which would give Syria to the boy twin after he came of age. Antony gave him yet more, including Armenia and Parthia, which would have come as a surprise to the still unconquered Parthians. To Cleopatra Selene he bestowed Cyrenaica and Libya. "Antony was coming forward as the direct representative of the Hellenistic kings," according to the Russian historian Dimitri Rostovtzeff, "and proving to Rome that the plan of shifting the center of the Roman Empire from Italy to the East was no novelty for him."[17]

Antony's fellow countrymen learned that he not only acknowledged three of Cleopatra's children as his own in public, but, without even notifying the senate, also had the unmitigated effrontery to parcel out Roman provinces and independent kingdoms to minors. They finally recognized that he and his Egyptian queen were uniting the East against Roman civilization. Octavian, who had taken the name Caesar Augustus (see fig. 1.6), declared that Antony had become a traitor to Roman ideas, thanks to the "Egyptian whore's" corrupting influence.

Marc Antony's behavior did, however, have a salubrious effect on Rome, although not as he intended. His actions reflected so poorly on the **Isiac** fraternities in Italy that their followers were popularly discredited and their conspiracy rendered impotent. And none too soon. According to Chanler:

> At the time Augustus began remodeling the Roman Republic, the greatest organization in the world had been for some time and still was the pontificate of the Alexandrine cult, the priesthood of Isis, and the Isiac societies. Like moles, they did their work in the dark, spreading their ramifications through every stratum of society. The discipline among them was iron; there were no disruptive conflicts for authority, nor civil wars within the brotherhood, such as one time seemed likely to split the secular world asunder.[18]

Cleopatra's aggressive intentions and manipulation of Antony, now her willing slave, had become obvious, and both sides prepared for war. She was proud of her foresight in building up the Egyptian fleet, which, combined with Antony's ships, numbered no fewer than eight hundred vessels. Many of these battleships were eight- or ten-banked galleys, larger than anything the Romans had afloat. And Antony's land army was huge. It comprised one hundred thousand infantry with twelve thousand cavalry. Against this formidable challenge, Caesar Augustus fielded an equal number of horses, but twenty thousand fewer foot soldiers. Worse, his warships numbered only 250 vessels, less than a third of Cleopatra's combined fleets. Because of his numerical inferiority, she insisted that Augustus be confronted by an all-out naval engagement, against the strategy of Antony and his generals.

Fig. 1.6. Caesar Augustus in the robes of chief priest (Capitoline Museum, Rome)

They argued that their land forces were additionally bolstered by the armies of eleven allies, from Bocchus of Mauretania, on the Atlantic shores of North Africa, to Herod and Sadalas of Thrace, today's Bulgaria. These powers would be held in readiness, she countermanded, to overrun Rome and the West after Caesar's defeat at sea. Acquiescing as always to her demands, Antony lay in wait with his fleet near Actium, allowing the Romans to reach striking distance. As the uneven engagement got under way, the numerically disadvantaged Roman warships proved to be more skillfully manned. They outmaneuvered the enemy's powerful but clumsy vessels to score the first sinkings and captures, trapping Antony's larger fleet inside the Gulf of Ambracia.

In September he attempted a full-scale breakout, resulting in the Battle of Actium. The Roman blockade bulged under the assault, and the fighting had not turned to either side when Cleopatra's flagship, the *Antonias*, christened after her browbeaten lover, abruptly hoisted sail and fled in the company of sixty warships. Her withdrawal astounded the Romans, who had so far inflicted little damage on the enemy. The men of Antony's fleet were paralyzed with disbelief, but their commander was not. He jumped ship into a galley to pursue Cleopatra, abandoning his command without a word of explanation. Those of his followers who witnessed such shameful proceedings lost all heart for the struggle and retired their ships from harm's way. Others, ignorant of these events, continued to fight on but were defeated piecemeal by Roman sailors and marines greatly encouraged by the queen's inexplicable retirement. As word of the outcome at Actium reached Antony's land forces and his expected appearance never materialized, the one-hundred-thousand-man army and its allies melted away, many officers and troops defecting to Augustus.

Cleopatra's whole life had been obsessed with world dominion. But when she saw for the first time the violent deaths of men in combat, the experience terrified her. She panicked and fled, and with her went all hopes for the fulfillment of Ptolemy's prophecy. Cleopatra had the heart for conquest, but not the nerves for it.

Catching up with her, Antony came aboard and crouched in the bow of the ship, hiding his face in his hands. The ruin that inevitably overwhelmed them both is among the most famous tragedies in history. But it would have been far more tragic had they succeeded in their ambition to overwhelm civilization and replace it with what was, even then, an outmoded tyranny.

2
Mauretania

To speak the names of the dead is to make them live again. It restores life to those who have vanished.
ROMAN ADAGE

With their mother's suicide, the fate of Cleopatra's four children was uncertain. As long as they lived, they threatened the world as potential figureheads around which ambitious men, acting behind the fame of their royal puppets, could rally the defeated but not extirpated forces of revolt. Such had already occurred just after Cleopatra's death, when Caesarion, her eldest son by Julius Caesar, was spirited away by court officials to Greece, where they began to set him up as their young cat's-paw and to rally Marc Antony's scattered followers. It's possible that the Roman authorities, having learned of the danger in time, had Caesarion assassinated and then swiftly liquidated the men who tried to use him on behalf of their own political motives. More likely, the unfortunate lad fell victim to lethal squabbles among various factions seeking to exploit his name.

Whatever the case, he died violently, while the dead queen's surviving children—Cleopatra Selene, Alexander Helios, and Ptolemy

Philadelphus—were taken captive to Rome. They were forced to march in the victory parade staged by Augustus to celebrate his successful campaign against Antony and Cleopatra. Cleopatra's nine-year-old twin daughter and son helped their little brother trudge sadly behind a colossal float depicting their dead mother in chains. The crowds did not jeer at their pitiable appearance, however, and some voices were raised against Caesar, who was criticized for the indecency of lording his triumph over innocent little children. But Augustus had no intention of doing them any harm. Far from it: He placed them in the gentle care of his much abused sister. For the next several years, Octavia nurtured them with kindness and understanding in the royal nursery at the palace on Mount Palatine. In addition to her compassionate attention, they benefited from the same kind of excellent education available to any prince or princess of the imperial family, and grew up to become highly cultivated young members of the Roman aristocracy.

Fig. 2.1. A statuette of Alexander Helios as a teenager at the time of his stay in Rome's imperial palace. The pyramidal crown he wears probably indicates some rank in a religious sect. (National Museum, Cherchel, Algeria)

True to his heritage, Alexander Helios showed an early interest in political science, and was provided with a special tutor adept in statecraft. This was Nicholas of Damascus, a Gentile and former high-ranking officer in Herod's army. With the onset of adolescence, Alexander Helios tended toward obesity, a genetic characteristic that plagued many male members of the inbred Ptolemies. As he matured, however, he grew more physically fit (fig. 2.1). Philadelphus became a champion in sports, particularly chariotry, excelling as an accomplished equestrian. But Cleopatra Selene grew into a great beauty famous for the long tresses of her pure blond hair that shone like polished gold in the sunlight (see fig. 2.2).

Another royal child orphaned by an earlier war had likewise been raised in the imperial nursery. This was Juba II. His name meant "commander," appropriately enough because he was descended from a long line of kings going back a thousand years to the early Phoenician occupation of North Africa. The Phoenicians were a Semitic-speaking people originally confined to the Near East. When the sudden end of the Bronze Age about 1200 B.C. shattered the old Homeric world order, the Phoenicians expanded as far as their mercantile instincts could take them, which was very far indeed. They inherited the maritime tradition of their Bronze Age predecessors, building exceptionally seaworthy ships and plying their trade throughout the Mediterranean, beyond the Pillars of Melchart (today known as the Straits of Gibraltar), into the Atlantic Ocean. The Phoenicians strung together a vast, expanding commercial empire through which the goods of the civilized world spread from Mesopotamia to Britain and Morocco.

Fig. 2.2. Cleopatra Selene, queen of Mauretania and daughter of Cleopatra the Great (National Museum, Cherchel, Algeria)

Around 600 B.C., they were engaged by the Egyptian pharaoh Nekau II to circumnavigate the African continent, an accomplishment unrivaled until the voyage of Portugal's Vasco da Gama two thousand years later. The Phoenicians built dozens of harbor cities, such as Berytus (Lebanon's Beirut), Tyre (cursed by Ezekiel in the Old Testament), Motya in Sicily, and, most famous of all, Kart-Hadasht, the "new town," erected on a triangular peninsula in the Gulf of Tunis. Carthage grew to dominate much of North Africa, including the area covered today by coastal Libya, where large-scale silver-mining operations generated exceptional wealth and influence. But the Carthaginians used their power to challenge Rome politically, and thus, during two Punic wars, they were vanquished, their great city finally razed to the ground in 146 B.C.

Before their defeat, they established trading centers farther west along the North African coast to Numidia, a barbaric land roughly corresponding to the boundaries of modern Algeria. Phoenician impact on the Numidians, particularly upon their regional leaders, who commonly intermarried with their Punic "betters," was culturally and genetically profound. According to Dr. Covey, "Carthage had civilized the nomadic Numidians."[1] Numidia, Latin for "nomad land," was composed of numerous, often contentious, unsettled tribal peoples of various racial origins. During the third century B.C. they were loosely united into a polyglot state by an early, ambitious king, Masinissa. First an ally of Carthage, he switched sides when Roman victory seemed imminent in the hope of picking up former Carthaginian territories to make himself the ruler of Africa. But before he could acquire any extra real estate, the Romans forced him to divide the kingdom among his three eldest heirs: Micipsa, for civic affairs; Mastanabal, for the treasury; and Gulussa, for the army.

The country proved nonetheless troublesome to Rome, with sometimes dangerous outbreaks of revolt, the most serious of which took place in 49 B.C. Instigated by Juba I (see fig. 2.3), it was thoroughly suppressed three years later at the Battle of Thapsus. His corps of war elephants had stampeded, throwing the Numidian infantry into disarray. After this crushing defeat, Juba sought refuge at Zama, near the sea, where he planned to join his wife and baby son. But he was turned out of the coastal city by his own countrymen, who were disenchanted by his failed ambitions. Disgraced, he retired to a royal pavilion some distance away, where he feasted on a magnificent banquet with his handful of remaining followers. After this last supper, he fought a duel to the death with his chief general so that both men might die honorably. As the mortally wounded victor, Juba ordered a slave to finish him off. Thereafter, Numidia ceased to exist. Julius Caesar, the victor at Thapsus, formed it into in a Roman province, Africa Nova, "New Africa."

Among the spoils of victory in 46 B.C. was the former Numidian crown prince Juba II. Like Cleopatra Selene years later, he was forced to adorn a Roman triumph dramatized to celebrate the destruction of his country and the death of his parent. The foreign child was carried in procession, accompanied by ranks of his fellow Numidians, who, in chains, cringed from the huge, noisy crowds of spectators. But at just two years of age, the little prince was too young to understand what was going on. To his toddler's imagination, the procession, with its colorful floats, wild animals, many thousands of people, and loud music, must have seemed like a spectacular parade with him as its focus of attention. He smiled ingenuously and waved with untiring glee at the Roman citizens, breaking and winning their hearts. Plutarch described him as "the happiest captive that ever was."[2]

Keen to the political implications of little Juba's sudden popularity, Caesar had him installed at the palace, where the lad was lavished with all the courtly attention of an aristocratic family member. He took to his education with enthusiasm, excelling as a serious student, and exhibited a natural cheerfulness that won him many friends. Respect for him broadened, as his total lack of political or financial ambition—in this, the most politically and financially ambitious city in the world—became generally known. Caius Julius Juba

Fig. 2.3. Juba I, the Numidian king, who committed suicide by engaging in hand-to-hand combat with his best friend (National Museum, Cherchel, Algeria)

(see fig. 2.4), as he was now known, developed into a true scholar whose chief motivation in life was the eternal quest for knowledge. He was truly happy and content only when studying, attending lectures, or writing. Fluent in Numidian, Latin, Greek, and Phoenician, he was something of a genius, publishing dozens of books on history, botany, geography, zoology, philosophy, poetry, grammar, painting, music, dance, sculpture, and travel. His work on the history of theater alone comprised seven volumes. Chanler writes:

> Although we know the titles of at least fifty-two books, probably there were more. His works (fragments of which alone remain) not only won the praise of his contemporaries, but were freely drawn upon by them and the writers of later generations. [Called a] compiler . . . by some, he followed the fashion of his day, and might better be called an encyclopaedist.[3]

Juba had personal friends among Rome's literary elite, including Virgil, Horace, Livy, Agrippa, and Horace. Ovid himself referred to Juba

Fig. 2.4. Caius Julius Juba, or Juba II (National Museum, Cherchel, Algeria)

as an *auspex,* the "best man" of letters, and Pliny said he was "one of the great scholars and writers of the age."[4]

With his auburn hair cut short, Juba II was not only Romanized, but also a fanatic proponent of the *Pax Romana,* Rome's civilizing mission in history. He regarded Roman civilization as the highest goal of mankind, symbolizing the triumph of ennobling ideas, legalized justice, and cooperative organization over the ignorance, despotism, and chaos endemic to the outside world. Juba did not despise his foreign origins, however. On the contrary, he firmly believed that the best parts of his Phoenician heritage from dead Carthage and Numidia could contribute to the cultural enrichment of the world through Roman literature, which was spreading to every corner of the empire.

Caesar Augustus much admired Juba, but was concerned that an excess of intellectualism so early in life was not good for the development of the young man's character. Still in his mid-twenties, he was, in effect, drafted into the army for two years. He rose to become an officer stationed in Spain, where he conducted operations against a rebellious Celtic tribe, the Cantabri. To the surprise of his many friends back home, he distinguished himself as a coolly calculating commander and administrator, as just and merciful as he was efficient, and, when circumstances justified it, ruthless. These classical Roman qualities did much to endear him to the native Iberians under his reasonable sway, and he established permanent personal contacts among them that would be helpful later in his career. Even then, he was noted in Spain for his cultural and scientific pursuits, interests that the occupied peoples found curiously amusing for a Roman officer.

Returning to the capital in 27 B.C., Juba was glad to get back to his scholarly work. He became a researcher in Greek manuscripts at Rome's first public library, the wonderfully named Hall of the Goddess of Liberty, the Atrium Libertatis, founded twelve years earlier by Asinius

Pollio. There Juba became assistant to the head librarian, Gaius Julius Hegenus, and, with Gaius's permission, selected certain highly literate Greek prisoners of war and organized them as educated copyists, the *servi literati,* at another library-museum, the splendid Temple of Apollo. Innovative actions such as these brought him to the attention of Marcus Terentius Varro, a poet, archaeologist, historian, and grammarian chosen by Caesar to dramatically expand the Roman library system. Together, the two scholars were responsible for an upsurge in general literacy through their establishment of new branches throughout the city.

Chanler writes that Juba had by this time become "a particular favorite of the Emperor," who began to consider the young man for bigger things.[5] He gave him back his country, the former Numidia, to rule as proconsul. Juba did not jump at the chance, however. His happiness was in the cultural life of Rome. Africa Nova, he knew, was a desert in more ways than one. Still, the assignment might offer him the opportunity to put his learning into effect somehow. While he was still undecided about the post, Empress Livia introduced him to the most beautiful girl he had ever seen. The fifteen-year-old Cleopatra Selene was no less taken with him, despite their thirteen-year age difference, and they were wed, to great popular acclaim, in 25 B.C. The couple sailed soon after for their new home near the Mediterranean shores of North Africa.

They found conditions there in sharp contrast to the civilized world left behind. The most basic necessities were in short supply, the climate hellish, and the people surly. Setting up house in the dilapidated palace at Cirta, the former Numidian capital, Cleopatra Selene had her work cut out for her, performing endless hours of manual labor just to get the place in livable shape. Juba began by carrying out an unprecedented survey of Africa Nova, from which resulted the first map of the province. Roads and aqueducts were of primary importance, and these he began building at once in the best tradition of Roman engineering, simultaneously solving the local unemployment problem. He studied a colossal irrigation canal built in Numidia by a very ancient pre-Phoenician people, the Garamantes. The ten-foot-high, twelve-foot-wide stone structure ran almost one thousand miles from the Ghat

Oasis to their capital city, Garama. Modeling his own aqueducts on this spectacular achievement from prehistory, Juba brought precious water and, with it, a revived agriculture to his homeland.

After two years, he and Cleopatra Selene had much to show for their exertions, but they were burning themselves out in an environment that did not suit them. They petitioned Augustus to return to Rome. Ever the man of political compromise, he refused their request but gave them another, more appealing realm to govern. In the spring of 23 B.C., the royal pair traveled sixty miles west along the gravel highway Juba had already built from Icosium, Numidia's first and largest city, to Iol. This was the dilapidated capital city of Mauretania. In area, the land roughly comprised all of present-day Morocco and part of Numidia, about eight hundred miles between the shores of the Atlantic Ocean and the Ampsaga River with an indefinite southern boundary bordering the black kingdoms of West Africa.

Its history was lost in myths of Hercules and Atlantis. According to Plato's early-fourth-century B.C. dialogue the *Kritias,* an Atlantean king, Autochthon, was the country's first ruler. The Atlas Mountains were supposed to have been named in honor of the emperor of the Atlanteans. Both the earlier Greek historian Herodotus and the later Greek geographer Diodorus Siculus wrote of a people, the Atlantioi, still living on the Mauretanian coast, who claimed descent from Atlantis. In the story of Antaeus, the giant son of Poseidon and Gaia, defeated by Hercules in Mauretania, some detected a mythic version of the historical Greek victory over the Atlanteans as described by Plato. Indeed, there were and still are pre-Phoenician ruins at Mogador (present-day Essaouira), where the remains of a gigantic harbor facility stand not far offshore; among the massive stone walls of Lixus; and at the site of the curiously Atlantean city of Volibulis, with its identifiably concentric pattern. Greek tradition told of Hectapylon, the "city of a hundred doors," built by Hercules during his world travels and located somewhere in western North Africa. Myth subsequently became reality when a circular megalithic site comprising one hundred upright monoliths resembling doors was indeed found by early-twentieth-century French archaeologists near the Moroccan village of Zora.

During the sixth and seventh centuries B.C., the Phoenicians built important cities along the Atlantic coast of North Africa, but they found these territories already occupied by a people known as the Mauri, who claimed to have originated in Asia Minor, where they had been allied with the Trojans. Their claim seems more than legendary, because the Greeks did indeed know them as the Maurusi, from the Anatolian Maurusia, in modern-day coastal Turkey. With the defeat and burning of Troy in 1240 B.C., the Mauri were forced to find refuge where they could. Hounded by Mycenaean Greeks and opposed by hostile natives at every turn, they fled out of the Mediterranean Sea, finally settling along the shores of the land subsequently named after themselves.

But they were neither alone nor the first inhabitants of this land. Several indigenous peoples, including the Berbers, Tuaregs, and Gaetuli, engaged them in constant strife. For centuries the Mauri were not only forced to defend themselves against these hostile natives, but were also incessantly victimized by bands of roving pirates who plundered the entire coastline of western North Africa. Over time, however, the Mauri asserted themselves, brought a modicum of order to this disorderly land, and formed a lasting if uneasy peace with the fractious natives. By 118 B.C., these disparate racial groups were cobbled together into a quasi-monarchy headed by Bocchus I. Although beset by perennial insurgency, he was able to develop Mauretania's previously neglected wheat agriculture, turning a handsome profit in export sales and thereby gaining some measure of international recognition.

His son, Bocchus II, dying without an heir, willed his country to the Romans. But they did not want it and set up their own choice, Bogund, a native Mauri, to rule Mauretania for them. It was a decision they would live to regret. As an example of his savage cruelty, he ordered one of his enemies, Magudulsa, buried in the ground up to his head, which was then trampled by an elephant. In Cleopatra's war against the West, Bogund joined ranks with Marc Antony, who foolishly gave him a command against Octavian at Methone, an island off the southwestern tip of Greece. Agrippa's rapid seizure of the strategically important island placed his numerically inferior Roman fleet in a position more favorable for attacking Antony's ships at Actium, where the campaign was decided. Bogund fled during the subsequent invasion, and was captured and executed by the Romans, who reduced his country to colonial status.

This was the defunct kingdom Cleopatra Selene and Caius Juba received as a questionable gift from Augustus and over which they were to rule conjointly as queen and king. The young couple immediately fell in love with the country's long coastline, hot sun cooled by ocean breezes, deserts, mountains, rivers, and exotic wildlife. They rose to the challenge of transforming the land into a civilized state where they could build their permanent home and raise a new dynasty of rulers.

Juba's first task was to win over the fiercely uncompliant peoples of his new realm. At a potentially dangerous mass meeting, he forthrightly promised them tribal autonomy within the framework of the new society he wanted to build. Their gods, customs, languages, and way of life would be honored, even protected. Disputes would be settled henceforward by the exercise of just laws. But banditry and outlawry by anyone would be ruthlessly punished. All he asked from his skeptical listeners was peace and, if they so inclined, cooperation. The Roman legion would stay on, not to hinder the tribes in any way, but to put an end to the bloody strife that had made a mockery of Mauretania in the eyes of the rest of the world. He was giving them a chance at civilization. They could take it or leave it.

Juba's transparent sincerity, almost naivete, caused the suspicious Tuaregs, Berbers, Gaetuli, even the disenfranchised Mauri and remnant Carthaginians to pause for sullen consideration. He was unlike any leader they had ever seen, offering them friendly peace with one hand and iron justice with the other. The mood of uncertainty was dispelled when the politically astute Juba, having ended his forthright speech to their grim silence, unexpectedly leapt upon his horse and gave them a superior demonstration of his equestrian skill. The tribesmen, deeply proud of their horsemanship, roared their appreciation. Here indeed was a real man to lead them![6]

There followed unprecedented construction and growth that hadn't before been witnessed in this part of the world. Cleopatra Selene and Juba strove with tireless enthusiasm to transform Mauretania into a modern society. Their inspiring leadership, Roman education, and hard work conjured a new kingdom from this imperial backwater. Juba began as he had in Numidia, this time building Africa's largest aqueduct, which ran from the distant Atlas Mountains to numerous cities and towns. To the empire's eighty thousand miles of roads he contributed many more, linking almost every part of his realm. These measures, combined with the improved agricultural practices he introduced, resulted in bumper wheat crops. The country's plenty was so great that exports were made to Italy and eased grain shortages in Egypt, while netting Juba handsome profits. Over time, Mauretanian wheat production continued to increase, gaining lucrative markets in Spain.

In his ongoing historical research, Juba learned that the Phoenicians had cultivated certain rare mollusks—the *Murex trunculus* and *Murex brandaris*—along North Africa's Atlantic coast for the production of a luxuriant purple dye used to color the kingly cloaks of only the wealthiest rulers. By studying the old Phoenician documents, he rediscovered how the one or two drops of liquid had to be expertly extracted from a small gland (the hypobranchial gland) in a living snail in order for the brightest hues to develop properly. First, the yellowish liquid was exposed to the sun, which caused it to darken. It was then heated carefully in a lead or tin pan—any other metal discolored the dye. After simmering for two weeks, the liquid was reduced to approximately one sixteenth of its original volume. This tedious process required some sixty thousand snails to produce a single pound of dye.

Because of the complexities involved in the manufacture of the purple stain, it sold for the equivalent of approximately twenty-eight thousand dollars per pound. The Phoenicians and now the Mauretanians became skilled dyers in the luxury export trade. They learned how to vary the tint through the admixture of other dyes. But their most expensive color was the lustrous imperial purple worn by the very wealthy. Regardless of their high prices, the production of Mauretanian dyes could hardly keep up with demand from Rome. Not only members of the imperial family, but also senators and rich citizens wanted to display the narrow purple band on their togas. The murex industry developed into a booming export business, and Juba built a purple-dye factory on a small island in the area of Autololes, on the Gaetulian coast. The tribal Gaetuli profited enormously by having the center of dye manufacture in their region, and they discarded their former hostility to outsiders.

Other luxury products exported by Mauretania included leather goods, ivory, olives, ostrich feathers, and doomed beasts for Rome's cruel circus, for elephants still ran wild in northwestern Africa. In fact, it was from here that Hannibal gathered his pachyderms for the invasion of Italy via the Alps. Other animals in demand from Mauretania were panthers, lions (King Bocchus had given lions to the Roman Sulla in 98 B.C.), asses, and camels, which had been introduced by Juba from his native Numidia, where his father deployed twenty-two camels in a kind of humped-backed cavalry at the Battle of Thapsus. Most important of all, however, were Mauretanian horses. They were small but much valued for their speed and toughness. Like their Trojan ancestors, renowned horse breakers of the Bronze Age, the Mauri excelled at breeding and training the animals.

Another successful export was entirely the invention of Juba II: His chemistry experiments resulted in the discovery of a new purgative, as extraordinarily effective as it was gentle. It was welcomed as a boon to medical science, and became known throughout the Roman world as the *Euphorbia regis jubae*. His innovative genius and hard work had opened the sluice gates of wealth, pouring abundance into his country for the first time. Its middle class burgeoned, the upper classes expanded, and Juba's treasury rapidly grew to become one of civilization's greatest troves, especially when gold was discovered and mined in the Atlas Mountains. As though all these riches were not enough, precious and semiprecious minerals were mined on a large scale, including high-grade copper, marble, rubies, and especially garnets.

Unlike some Eastern potentates, Juba reinvested his vast revenues in social and cultural development. To further capitalize on his prosperous grain production, he built a large commercial fleet, then protected it from piracy with a powerful navy known as the Augustus Alexandrina, after the predominantly Alexandrian crews that manned its squadrons. Their shipmates were a mix of Syrians, Spaniards, and Mauri, who, together with their former pirate enemies, had been superb, far-ranging seafarers for centuries. They sailed in a mighty complement of 151 war vessels divided into two classes—the battle cruisers in the Classis Alexandrina and the smaller faster ships in the Classis Syriaca. True to his Phoenician origins, Juba was determined to build Mauretania into a maritime power.

Accordingly, he turned the Numidian Cirta Rusicade into a port city, and opened a great harbor with two ports at Iol, the Mauretanian capital. He soon fell into a building frenzy, modernizing the ancient city of Volibulis, with its concentric layout reminiscent of Plato's Atlantis. He founded or built upon other metropolitan centers, including Chullu, Ruspina, Lixus, Tarmuda (today's Tetuan), and Leptis Minor. According to Chanler, "The urge for building was communicated to the other African cities, and soon all were being ornamented in Hellenistic style."[7] She concurred in this with Theodor Mommsen, Germany's great nineteenth-century scholar still regarded as a foremost authority on imperial Rome, who observed:

> The prosperity which now subsisted in Nova Africa is clearly attested by the ruins of its numerous towns, which everywhere exhibit baths, theaters, triumphal arches, gorgeous tombs, and generally building of luxury of all kinds, often excessive in magnificence. Not quite in the villas of the superior nobility, but in the middle class of the farming burgesses must the economic strength of these regions have lain.[8]

But Juba lavished most of his attention on the old city of Iol. Its name meant "return of the sun," perhaps a reference to its Phoenician builders and their solar religious cult. He was joined if not exceeded in his efforts by Cleopatra Selene, whose personal impact on the city became increasingly pronounced with time. Together they raised a new capital on its ancient foundations. Again, Atlantis-like, Iol was divided by three concentric walls, the largest enclosing one thousand square acres. Although the great city's monumental architecture became unmistakably Roman, its overall plan was patterned after Alexandria. But it was no copy, and took on the character of both its sovereigns and the local environment. Within its walls were a triumphal arch, a magnificent forum, and a grand basilica (or law court). The streets were laid out in avenues, boulevards, and alleys, and a processional, winding among numerous mansions, was adorned by hundreds of statues—colossal, lifesize, and small—from around the Mediterranean world.

The fashionable quarter lay west of an impressive outdoor amphitheater, indoor theater building, and an oval circus, in size almost as large as Rome's Circus Maximus, where all manner of spectacles took place. As Chanler points out, "Only Carthage at its height boasted three such edifices."[9] Nearby villas were attractive affairs of yellow and rose marble rearing among shade trees, irrigation canals, flower beds, fishponds, and water lily pools. As Covey writes, "Archaeologists were astonished at Iol's luxurious modernity."[10] Its two hundred thousand residents enjoyed several libraries and *thermae,* or huge public baths that were located in the southeastern section of the city. There were several theaters, the largest fronted by a larger-than-life-size statue of Caesar Augustus in full dress armor. "One cannot doubt," writes the French archaeologist Maxim Gsell, "that the capital of Mauretania possessed a place of scenic spectacles from the time of Juba, who called Greeks to his city as actors and wrote a history of the theater."[11]

Iol also boasted a fine public hospital in the Temple of Aesculapius, who was a mortal Greek doctor deified for his medical services to mankind. There were olive groves, public parks with marble fountains, a zoo, vineyards, and flower gardens. Gardens of other kinds were cultivated in and beyond the city where grew almonds, figs, pomegranates, pears, quince, lemons, and melons. Favored vegetables were peas, radishes, beans, artichokes, wild asparagus, garlic, onions, African truffles, and *cimmin* (possibly cinnamon).

A temple to Neptune, complete with a white marble colossus of the god, stood on cliffs overlooking the sea, and the harbor was always filled with merchant ships from Spain, Gaul, Italy, Egypt, Syria, and even Britain. Other Roman gods were venerated in their own sacred structures, including sky father Jupiter; earth mother Ceres; Pluto of the Underworld; Mercury, the divine patron of civilization; Mars, the war god; and especially Saturn, the heavenly champion of agriculture and protector of the soil. The Greek god of spiritual and artistic enlightenment, Apollo, the Carthaginian Tanit, the Persian Mithra, and the Egyptian Isis each had his or her own temple in Iol, as did the gods worshiped by the indigenous people, of which much less is known. They were described by the Romans as Diana Augusta Maurum, a Mauretanian version of the lunar goddess of mysticism, together with the more obscure *deus sanctus* Aulisua and *genius summus* Thasuni. At the summit of a high hill overlooking the city were temples to Bacchus, the god of prosperity, and Hercules, in which Juba venerated the spirits of his ancestors in the Heraclidae dynasty. In a remarkably short time, the capital had become, in Mommsen's words, "the residence of a cultivated and luxurious court, and a seat of sea-faring and traffic."[12]

The palace was a large, splendid affair of Carrara marble from Italy, local rose marble, pale white alabaster from the Upper Nile Valley, and rare, fragrant woods. Around a central, marble-paved court of fountains and statues were the royal apartments, each with its small, private garden bordered by porticoes. Many thousands of brilliant stones composed mosaics depicting mostly idealized agricultural and mythological themes across the floors and up the sides of some walls. A marble portico of green diorite columns surmounted by white capitals was more Alexandrian than Roman, as were the green porphyry and serpentine marble pillars that stood throughout the palace. The impressive royal residence was surrounded by oversized topiary figures of native animals, particularly elephants, and stands of imported Italian umbrella pines and cypresses from Crete.

The queen celebrated her annual Festival of Selene, honoring a Greek variant of the moon goddess, in a special shrine at the tip of Lochia Point, which jutted out from Iol into the sea. She also ordered the construction of a mausoleum for herself and Juba, which was built some miles from the capital, near the village of Tipaza, in a lonely region of the Sahel Mountains (see fig. 2.5).

Unusually circular, this flattened conical structure resembling a gargantuan beehive was faced with marble, sheeted in gleaming bronze, and surrounded by massive columns. Facing each of the four cardinal directions were what appeared to be monumental portals, but, fashioned

Fig. 2.5. Ruins of the mausoleum built by Queen Cleopatra Selene for herself and her king in the late first century B.C.

in the Egyptian style, they were actually false doors sculpted into solid masonry. The false door was a religious element, a gesture to the *ka,* or ghost of the deceased, which, being immaterial, could easily pass through stone into the interior. According to A. MacCallum Scott, an early-twentieth-century authority on Barbary, as northwest Africa was known then:

> [T]he monument, which is circular in form, stands upon a square platform. Its base is like a huge drum, 36.5 feet high and 198 feet in diameter. The circumference is divided into sixty equal spaces by sixty Ionic columns attached to the wall and surmounted by a frieze and cornice. The upper part of the monument is in appearance like a blunt cone, or rounded pyramid. It rises in a series of high steps some seventy-five feet above the cornice. The total height is about 110 feet, and it may originally have been as much as 130 feet from base to summit.[13]

The sepulcher was oriented to the rising of the sun, as part of an eternal rebirth principle in the mysteries of Isis. Shrubs and flowers near sculpted marble benches decorated the grounds. The monument's voluminous interior comprised a spiral labyrinth, at the center of which were sarcophagi for the king and queen. Frescoes depicting happy domestic scenes ran around the curving walls, which were additionally hung with household items the departed used in life. Anyone wishing to pay respects to the dead poured offerings of wine or milk into the mouth of a long lead tube that extended down through a receptacle inside the tomb. Scott recounted: "Far from the Nile, it nevertheless, by its shape and by its colossal proportions, recalls the pyramids."[14]

Even though its completion coincided with imperial Rome's early architectural florescence, the atypical structure attracted attention from the outside world. Pomponius Mela, a well-known author, lavished praise on the *monumentum commune regiae gentis,* the "public memorial of the royal family," in his *Du Situ Orbis*. Hundreds of years later, with the fall of classical civilization, it was remembered by Arabs as the Kubr-er-Roumia, the Tomb of the Romans, even though neither Juba nor Cleopatra Selene was Roman. Grave robbers hoping to lay their hands on Mauretania's ancient exchequer were lured to the site by local tales of the rulers' lost treasure. "A rank growth of legend, as well as of plants, clings to the monument," Scott observed.[15]

A typical story surrounding it told of a former slave who burned a magic papyrus inside the tomb, "and a stream of gold and silver pieces issued forth." He tried to capture some of the flying treasure in his unfolded burnoose, but "the spell was broken, the stream of money stopped, and the tomb closed fast again."[16] Such tales were fabulous echoes of the very real trove that had been sealed in the building with the mummified bodies of Mauretania's king and queen. "For nearly two thousand years," reported Scott, "the legend of buried treasure attached to the tomb, and many were the efforts made to wrest it from its rocky strong room. Generation after generation of would-be plunderers expended their strength in vain, trying to force the secret."[17] The coffined bodies of the royal pair for whom the mausoleum was raised had long before disappeared.

During 1825, the structure was further damaged in an earthquake, and again, forty-two years later, by warships of the Algerian navy, whose gunners used it for target practice. Despite these depredations, a stone shell of the royal tomb, long ago gutted and stripped of most exterior refinements, still stands at the western edge of the Metija plain extending into modern Algiers.

Scott lamented:

> Caesarea became known as the Athens of the West. All has perished. The looted pillars adorn many an Arab mosque, and some relics of statuary and mosaic may be found in the museums of Algiers and Paris. Corn and vines grow upon the site. For miles around, the plowshare turns up in every field fragments of statues, columns, capitals, inscribed stones, and exquisitely carved marbles. What treasures still lie underground we can but guess.[18]

3

SPQR: For the Senate and People of Rome

History is little more than a register of the crimes, follies, and misfortunes of mankind.
EDWARD GIBBON, THE RISE AND
FALL OF THE ROMAN EMPIRE

For three years after Juba and his teenage bride left Rome in 23 B.C., Augustus kept receiving reports of the miracle rising in the wastes of North Africa. Regretfully preoccupied with military and political affairs in Spain, he was unable to visit Iol. But the emperor was deeply proud of his adopted children and showed his admiration for their productive efforts by renaming the Mauretanian capital after himself, thereby elevating the city to imperial status. Henceforward, Iol was to be known as Caesarea. And that was not all. The senate unanimously voted Juba the rare title of *regius amicus,* an official "royal friend" of Rome.

These awards were not merely diplomatic gestures, but also carried enormous, almost unprecedented political and commercial weight throughout the empire. Wealthy Romans now flocked to Juba's kingdom, and not just as tourists. Many came to stay, vacationing or retiring in opulent seaside mansions. Others sought to invest or speculate in the country's abundant grain production. Over time, an affluent Roman community flourished in Caesarea and along the coast of the Atlantic Ocean. Romans settled in the luxury seaside resort towns of Igilgili, Saldae, Rusazu, Rusguniae, Gunugi, and Cartenna. Farther inland, Zuccabar was a retirement community for military veterans of the *Augustan* and *Oppidum Novum*. After only a few years of hard toil, Mauretania had gone from a colonial backwater to the first of senatorial provinces. Later, Plutarch was to describe Juba II as "the most accomplished of monarchs," Plato's ideal philosopher-king, a man bold in the arts of peace.[1]

So impressed was Caesar Augustus with Juba's achievement that he eventually elevated the young man's North African experiment to semi-independence, hinting it would eventually emerge as its own state and Rome's best ally. In giving Mauretania this status, some of the burden of running the already overextended imperial enterprise would be lifted from the emperor and his successors. As an important step in that direction and an expression of his hopes for Mauretania, he allowed the two regents to mint and issue their own coinage. In evidence of their shared power, Juba's profile often appeared on one side of these coins, his wife's on the reverse. Cleopatra Selene was, in fact, the first queen in the Roman Empire to produce her own coins. Her husband also felt free enough to pair himself with his family's divine ancestor Hercules, whose image also figured on many coins issued by the Mauretanian mint. Examples have been found in modern times as far away as the Balkans, in Croatia, testimony to the extensive commercial network operated by Juba.

Caesar's timing for these important honors was not without significance. They were bestowed as a birthday present to Cleopatra Selene's first child, a son born at the former Iol, now Caesarea, in 20 B.C. The next year, Augustus dedicated a special gold coin to the boy on his first birthday, an unprecedented gift—only the emperor could issue gold coins—and therefore a token of the high esteem in which he held the baby's mother.

The next sixty years represented a period of peace and prosperity, as Mauretania became an important part of the worldwide rule of Roman justice and order. Juba happily relinquished many of his political responsibilities to the queen, who acted with real authority in Caesarea, especially when the king was away on his prolonged travels to distant lands such as Arabia. In 4 A.D. he moved on to Cappadocia, where he befriended King Archalaus. These were mostly scientific and cultural research expeditions, for he was at heart more scholar and author than ruler.

However much his lengthy absences from home may have benefited the worlds of science and art, they contributed to the neglect of his son and heir, Ptolemy (fig. 3.1), who was in need of the special upbringing only a father who was also the monarch could provide. Young Ptolemy's political views were not being formed by his Romanized father. Instead they were nurtured by Cleopatra Selene, who was infected with the same Hellenistic dream of world conquest that led to the undoing of her own parents. It would some day prove no less perilous for her only son.

Ostensibly, he had been named after Cleopatra Selene's younger brother, Ptolemy Philadelphus, the recent victim of a fatal accident—the eighteen-year-old charioteer was killed during a race at Rome's famed Circus Maximus. But some speculated that the queen had actually named her son as the hoped-for embodiment of the old Ptolemaic prophecy, just as she herself was named after the ambitious Cleopatra the Great. As though daring to underscore the relationship, just when her son was born, Cleopatra Selene issued a coin portraying the likeness and carrying the name of her still-reviled mother. She even minted coins bearing the image of a crocodile and an elephant, traditional emblems, respectively, of the Ptolemaic dynasty and Cleopatra VII. Other coins she issued favored symbols of the Isis cult. As Chanler notices, "One senses her always pushing into the background what is foreign to her culture and birth. An assertion of this is found in her coinage, which lays such stress on her Ptolemy heritage and her divine genealogy."[2] Like her mother, she issued separate coins, not in Latin, as did her husband, but in Greek—a subtle challenge to Rome, however slight, that went mostly unnoticed at the time.

Fig. 3.1. Prince Ptolemy of Mauretania (National Museum, Cherchel, Algeria)

"Juba II and Cleopatra Selene," writes Covey, "in naming their son Ptolemy, were conscious of inheriting the Ptolemaic tradition."[3] At her wedding in Rome, Crinagoras, her Greek tutor, had recited a tongue-in-cheek epigram in which he extolled her marriage to the North African king as "uniting Egypt and Libya into one vast kingdom."[4] As Chanler further points out, "Cleopatra Selene never forgot her glorious ancestry."[5] Doubtless, her son was never allowed to forget that his grandparents were Cleopatra the Great and Marc Antony, and a fatal sense of destiny may have been implicit in that recognition. Cleopatra Selene also imported the Isiac cult, with all its ornate personnel and theatrics, into Mauretania, where, through her sponsorship, it pushed aside all other religious beliefs to become the realm's prominent theology. Just as her vainglorious mother did before her, Cleopatra Selene participated in numerous public ceremonies dressed as Isis.

Because of its maritime nature, Mauretania was a cosmopolitan kingdom. But the queen went further. Her declaration of total religious freedom attracted an influx of foreigners from around the Mediterranean world and the Near East. A Jewish community sprang up in Caesarea, and early Christians, despised everywhere else for their intolerance of other faiths, eventually found sole refuge there. They were accompanied by bizarre representatives of numerous fringe cults that were beginning to proliferate throughout the empire. Many were patently demonic, such as the Bacchantes, whose demented behavior led to their justifiable ban from Rome. At the same time, during the early first century A.D., Gnostics, who were despised alike by Romans and Christians for their *gnosis,* or secret knowledge, which alone was supposed to liberate the human spirit from the material world, were permitted to exercise their arcane rituals in Mauretania.

Like her mother, Cleopatra Selene envisioned the Isiac cult as a theological matrix that would meld and lead all other religions and diverse peoples of the world in a global enterprise with her at the top, because she was, after all, the living Isis. The world was spared whatever this dangerous megalomania may have amounted to when the queen died suddenly of apparently natural causes in 6 A.D. at forty years of age. Cleopatra Selene's passing during an eclipse of the moon, the heavenly body that was her divine namesake, was a meaningful coincidence the woman's mourners regarded as confirmation of her godhood.

Her death at least had a salubrious effect on Juba. He put an end to his travels, took charge of his son, and, as the most effective way to educate Ptolemy for the complex duties of kingship, shared the regency with him. Beginning in 9 A.D., the king minted coins bearing the legend "Ptolemeus Regis Iubae Filius," thereby announcing to the world that monarch and prince were conjointly in charge of the realm. For the next eighteen years they ruled wisely and productively. In an effort to undo some of the dangerous notions instilled by his wife before her death, Juba cautioned the young prince, "Work with the Romans, and they will work with you. Fight them, and you fight alone."[6]

Mauretania's father-son reign was marred only by outbreaks of serious banditry along the Numidian border. Seeking to take advantage of what they imagined would be Roman internal dissension after the death of Augustus in 14 A.D., the bandits' cutthroat chief, Tacfarnias, united a number of rebellious tribes through a combination of resistance rhetoric and promised plunder. He was a Roman army veteran but had deserted and returned to Africa, where he made a name for himself as the head of a band of robbers. His bravado appealed especially to the numerous but leaderless Musulamii residing along the southern slopes of the Aures Mountains, where he sometimes headquartered. Tacfarnias was likewise proficient in the hit-and-run tactics of guerrilla warfare, and led devastating strikes deep into Mauretania itself, pillaging the countryside, terrorizing defenseless inhabitants, and laying waste an important commercial city, Auzia. Following this success, he became something of a Numidian Robin Hood to native populations living outside the centers of civilization, with his enticing mix of looting and tribalism.

According to Mommsen:

> The insurrection extended eastwards as far as the Cinithii on the Little Syrtis and the Garamantes in Fezzan, westwards over a great part of Mauretania, and became dangerous through the fact that Tacfarnias equipped a portion of his men after the Roman fashion on foot and on horseback, and provided them Roman training; these

gave steadiness to the light bands of the insurgents, and rendered possible regular combats and sieges.[7]

He so outmaneuvered Mauretanian forces that Juba was forced to ask for help from the Roman governor of Numidia, Cossus Cornelius Lentulus. Together with Proconsul Furius Camillus, the Third Augusta Legion wiped out most of the marauders, killing their second most important leader, Mazippa. Tacfarnias himself escaped with a handful of followers. Ever the resourceful ringleader, he was soon at the head of another rebel army rampaging across eastern Africa Nova with elusive impunity.

Prince Ptolemy accompanied all the anti-bandit campaigns, learning much from the Romans. He lured Tacfarnias back to a rebuilt Auzia, where the brigand had enjoyed his greatest achievement. This time, however, the robber bands were utterly crushed between the legionary pincers of Ptolemy and P. Cornelius Dolabella, Rome's new military representative in Africa. As before, Tacfarnias eluded capture, but, with all hope gone, he committed suicide.

A grateful Roman senate dispatched one of its members to present Ptolemy with the *curulis,* the luxurious idealization of a military camp stool on which only the highest civil officers were privileged to sit; an ivory victory scepter; and the *toga picta,* a purple robe of supreme royalty decorated with golden stars. This last item was to play a fateful role in the destiny of Mauretania and its last king.

In another unique development, the senate awarded Ptolemy the title of Rex, Socius et Amicus—King, Fellow Associate, and Friend—in 24 A.D., which had been a year of triumph and tragedy. Juba, recently dead at age seventy-two, had not lived long enough to see his son's victory at Auzia and the honors that he subsequently received. Other accolades were to follow. Next, King Ptolemy of Mauretania (see fig. 3.2) was offered the honorary magisterial office of duumvir by the Spanish senate of Carthago Novo, and his statue was set up beside another of his father at the Gymnasium in Athens. Other Greeks in the Lycian federal assembly voted unanimously to raise their own statue of the new ruler.

While victory at Auzia brought Ptolemy far-flung acclaim, it likewise underscored Mauretania's continued reliance on Rome for protection. The independence hinted at by Augustus could not be granted by his successor, Tiberius, however favorably he too was disposed toward Mauretanian aspirations. As a young man, the new emperor had experienced an unrequited love affair with the golden-haired Cleopatra Selene, and came to respect her son. But as long as peace could not be maintained without the presence of Roman legions, they would remain in North Africa.

Fig. 3.2. Ptolemy as king of Mauretania (National Museum, Cherchel, Algeria)

Indeed, rebellions arose periodically throughout Ptolemy's reign. He tried to suppress these outbreaks with his own infantry, sometimes successfully, although rarely without Rome's regional governor to back him up. Ptolemy saw that the basis for Mauretanian independence must be a strong armed forces. Accordingly, he initiated a buildup aimed at enabling his countrymen to stand on their own feet against insurrection. His reforms were closely patterned after imperial models, and in this he was enthusiastically assisted by retired officers residing at Caesarea's Roman community. But in his determination to construct a military foundation for political independence, he did not neglect the rest of his royal duties. Over the eleven years since the defeat of Tacfarnias, Ptolemy maintained the orderly kingdom inherited from his industrious parents, and saw capably to its continued well-being.

4
Caligula: A Mind Abused

*Hateful is the power and pitiable is the life of those
who wish to be feared rather than loved.*

CORNELIUS NEPOS

While King Ptolemy was struggling to preserve peace and prosperity in Mauretania, the international status quo that was the *Pax Romana* seemed to be a permanent state of affairs. With the death of Tiberius in 37 A.D., however, prospects for change, if only within the imperial system, appeared imminent. He was succeeded by the twenty-five-year-old Gaius Caesar (see fig. 4.1), better remembered for the affectionate diminutive by which he was known to his father's soldiers when a child: Caligula—"Little Shoes," or "Little Boots," a name he naturally shunned. He was the first, although by no means the last, historical figure whose reputation was so maligned by his enemies that even today's generally accepted image of the man bears little resemblance to the truth.

Contrary to later events, Caligula was declared emperor on a great wave of personal popularity that was to last for the first half-year of his reign. His honeymoon with the senate ended only when he inexplicably fell into a deep coma, which he was not expected to survive. Reaction in some official quarters to his surprising recovery was less than jubilant, however. Close colleagues and senators, including members of his own family, had taken advantage of his debilitating illness to scheme for power and position, going so far as to malign his character. This revelation came as something of a shock to him, and his capacity for trust dwindled in proportion to growing cynicism.

Fig. 4.1. Roman emperor Gaius Caligula (Capitoline Museum, Rome)

He came to deeply detest the senate, not so much for the disloyalty of some of its members—that was to be expected—but because it was obvious to him and to anyone else who had eyes to see that the formerly estimable body of people's representatives had degenerated into a rich men's club of aristocrats whose political agendas had more to do with inflating their own wealth than that of the empire. Caligula hoped eventually to dispense with this plutocratic cabal, rather than to restore it to its republican ideals. Aware of their young emperor's growing hostility, the status-seeking senators fawned on him, seeking to curry his favor. He despised them all the more for their servility.

Caligula understood that the senatorial system had to be replaced. Some reformers suggested a return to the republic. But it had been badly discredited by the same kind of money men who now occupied the senate. As Covey points out, "The system was rigged so that the upper classes could always out-vote everyone else."[1] Moreover, Roman republican ideals, while barely practical on a limited scale, were open to demagoguery and corruption when applied to any numerically significant populace. Government had become unwieldy and bloated. By the time Augustus ascended the throne, there were more than a thousand senators. In his streamlining efforts to hedge in the growing power of plutocracy, he reduced their number to six hundred, but that was still too many. The eminent American historian Hutton Webster observed that the demise of the Roman republic deserved no regret: "This was not a loss to the world. Rome was no longer a small city-state to be ruled by mass meetings in the Forum. A gathering made up of the rabble of Rome, ready to sell their votes to the highest bidder, was scarcely a body fitted to represent the Roman Empire. The Senate likewise decayed."[2]

But Caligula's solution was the wrong one. Like his own great-grandfather Marc Antony, he had been infected by the Ptolemaic prophecy of a world dictatorship with him, as god, in control. This notion was instilled in Caligula at an early age. He had been brought up in the home of Antonia, Marc Antony's daughter. There she presided over a vengeful court where subversive power mongers, foreign and domestic, friendly to her father's Eastern dream, met to commiserate. Among them was Julius Agrippa, who became Caligula's closest lifelong friend and most powerful influence. Wealthy, well traveled, and in possession of a caustic wit, the Jewish potentate was a persuasive advocate of unbridled absolutism as the wave of the future. He flattered the impressionable young Gaius, assuring him his destiny was to rule the world unhindered by the pretense of a senate, with only he, Agrippa, at his side as his sole trusted confidant and "adviser."

Although the young emperor continued to show sincere solicitude toward the Roman people, who never abandoned their loyalty to him, he grew increasingly contemptuous of the senate. With body and mind weakened by illness, and with trust undermined by senatorial insincerity, Caligula's personality rapidly deteriorated into outrageous self-indulgence. He forbade any celebrations on the anniversary of the Battle at Actium, in which Marc Antony had been defeated. Agrippa urged him to go further, and Caligula authorized his own deification. Although the practice of pronouncing the godhood of a living king was common throughout the East, it was an affront to heaven the Romans condemned as hubris of the most shameful kind. But to Caligula, who spurned all other gods, his self-elevation to divinity was part of his plan to rule without any human interference. To show he was serious, he erected a pair of temples to himself, one of which he paid for with his own money. The other the senate was forced to fund.

Naturally, the emperor's megalomanical behavior generated powerful enemies in both the government and military. His secret agents had their work cut out for them, keeping track of all the numerous and growing conspiracies aimed at overthrowing him. A particularly serious attempt was made by the army commander of the Upper Rhine, in September 39 A.D. Caligula got wind of this intention to stir one of the empire's most powerful provincial garrisons in revolt, and hurried to the north accompanied by a large contingent of Praetorian guards.

The conspiratorial commander, surprised by the unexpected appearance of his emperor, was unable to destroy his secret papers, which then fell into the hands of the imperial police. The documents proved that the plot against Caligula was not limited to a few disgruntled officers. A list of names from the senate and army, including those of foreign allies, revealed a plan of unexpected proportions. Some of the people mentioned were leading figures of the time. Worse, the list implied that many more individuals were involved, but to what degree could not be determined. To survive, Caligula had to eliminate every one of them. But if he moved too quickly, the conspiracy might go deeper underground, rendering his task of rounding up all the participants much more difficult.

Rome seemed to him now too dangerous a place in which to remain. Instead of returning home, as everyone expected, he went south through Gaul to Lugdunum, the French city of Lyon. There he began implementing his anti-conspiracy offensive under cover of army maneuvers and public entertainments. His first unobtrusive move was to convene a meeting of various kings from Syria and other parts of the world. When they arrived in Rome, he sent word that he was on his way to meet them and begged their indulgence. Meanwhile, he had them spied upon by his undercover agents. The police did indeed confirm that at least some of the foreign monarchs were involved in subversive activities. Among them was the Mauretanian king, Ptolemy, who was already under suspicion for Juba II's friendship with the father of one of the recent military conspirators at the Upper Rhine. Now Ptolemy had been overheard in conversation with Marcus Junius Sitlanus, the proconsul of North Africa and the emperor's own father-in-law, who promised the monarch full independence for his country if he supported the plot to overthrow Caligula. Ptolemy eagerly complied, agreeing to put his strong fleet at the disposal of the conspirators.

While engaged in these dangerous conversations with Sitlanus, he received a personal invitation from the emperor to review the legions and enjoy the outdoor theater at Lugdunum. Suspecting nothing, Ptolemy set out from Rome to Gaul. Upon his arrival some weeks later, he was embraced by Caligula in front of the army command as the emperor's "dear cousin," and was shown every imperial courtesy. Both men, standing side by side as equals, watched troop maneuvers and thereafter shared at the banquet table.

Following several days and nights of royal camaraderie, they attended a stage performance together. To celebrate the occasion, Ptolemy appeared in public wearing his magnificent robe dyed with the deep purple tincture for which Mauretania was renowned. To the horrified astonishment of the capacity audience, Caligula flew into a hysterical rage, shrieking at the top of his lungs that Ptolemy had publicly insulted him by breaking the sacred convention of the imperial insignia: As a personal badge of office, only the emperor of Rome was allowed to wear purple robes. Anyone else who appeared thus attired was clearly attempting to usurp the throne, and must be condemned as a traitor.

As the king of the country that manufactured purple robes for every ruler in the Roman world, Ptolemy believed he was entitled to wear them whenever and wherever he chose. His appearance at the Lugdunum theater had not been a calculated insult, and other royalty commonly paraded in purple throughout Rome. True, Augustus had forbidden the wearing of purple by all save the emperor, but this restriction had been later eased by Tiberius. Moreover, the *toga picta* Caligula found intolerably offensive had been presented to Ptolemy sixteen years earlier by the Roman senate for his defeat of the Numidian rebel Tacfarnias. Coins minted at the time depict Ptolemy with the senatorial gifts he received, including the purple robe. In truth, accusing him of treason in this manner was Caligula's method of detaining the Mauretanian monarch without calling attention to the conspiracy the emperor was intent on destroying.

The arrest of a foreign king was big news. After the stunned and demoralized Ptolemy was transported in chains to Rome, he was visited in his prison cell by some of his influential supporters, including

Seneca, the Roman philosopher, dramatist, and statesman. Another friend, Sitlanus, had also been arrested and was later forced to commit suicide. Caligula's purge was on. Meanwhile, Ptolemy waited behind bars for his day in court, when he would plead his innocence and beg the emperor's pardon for having inadvertently offended him.

But his fellow countrymen did not wait. When word reached them of their sovereign's arrest, Mauretania's citizens erupted in revolt. Some went too far. They attacked their country's Roman communities, burning all the buildings to the ground and killing every man, woman, and child they could lay their hands on. Outrage swept through the people of Rome. They demanded revenge. For once, the senate and Caligula were in accord. They declared war on Mauretania, simultaneously ordering the execution of its monarch as an enemy alien. And so, in his sixtieth year, King Ptolemy, the empire's onetime Rex Socius et Amicus, died on the gallows.

5
Claudius: The Failed Peacemaker

*Cursed greed of gold, what crimes your
tyrannical power has caused!*

VIRGIL

A disturbing peculiarity that has apparently been missed by historians of ancient Mauretania was Cleopatra Selene's seeming inability to bear more than one child. Offspring are fundamentally essential to any monarchy, especially for a new dynasty trying to establish itself in a strange land. She did not give birth until three years into her marriage, and after this, no more children were forthcoming. The royal house of Mauretania comprised just three individuals, a very precarious state of affairs, especially for the future.

Curiously, Ptolemy never married and no children were attributed to him. His highly unusual bachelorhood was undoubtedly inexplicable and worrisome for his subjects. Cleopatra Selene's apparent difficulty to conceive and her only son's lack of offspring suggest both rulers likely had important difficulties with fertility. It is quite possible that Ptolemy, having fathered no known heirs, was completely infertile. Their problem may have been the inheritance of a house. The Ptolemies were plagued with all manner of recurrent physical and mental abnormalities after too many generations of inbreeding. While jealous to preserve their royal bloodline, they sabotaged their own genetic heritage. It's likely that among the recurring aberrations they encountered were difficulties with fertility.

Some kind of problem with Ptolemy may have been suspected from his birth. During that same year, 20 B.C., Cleopatra Selene's twin brother suddenly left Rome, where he had been in residence at the imperial palace since his capture as a child more than ten years earlier. His arrival in Caesarea is suggested by two commemorative coins minted in Mauretania and featuring representations of the sun with the moon. Although neither coin carries the name of brother or sister, they indicate to some historians (de la Blanchere, Eckhel, Merivale, Mueller, etc.) that Alexander Helios joined Cleopatra Selene at this time. Their implication is supported by a Mauretanian pendant of the same period displaying images of the sun and moon side by side.

Avuncular interest in her new baby may not have been Alexander Helios's only motive for undertaking the long journey, however. Might Cleopatra Selene have summoned her brother to stand in for Ptolemy in the event that her husband should die and her son be prevented from assuming the throne? Legally, Alexander Helios had been next in line after his sister to rule Mauretania. But he never showed any interest in North Africa, and doubtless preferred to remain in Rome enjoying the good life, as did his younger brother, Ptolemy Philadelphus. Politically unambitious, he was a safe backup, a usefulness that became more apparent with time.

Long before the death of Juba II, it seemed clear that his son would produce no heir. There was no one to succeed him but his uncle. It seemed, then, that with Cleopatra the Great's grandson the lineage of the Ptolemaic kings would die. And Alexander Helios had neither wife nor children. After him, if he did not die before his nephew, fifteen years his junior, what must become of the crown? The rule of Mauretania would be up for grabs, and that usually meant civil war, not a legacy worth considering. If anything should happen to Ptolemy, at least someone from the royal family would preserve the dynasty—so one contemporary line of reasoning must have gone.

When invasion came after King Ptolemy's execution in 39 A.D., Alexander Helios did not direct the Mauretanian armed forces. They were commanded by a freedman, Aedemon, probably because Alexander Helios, at seventy-five years of age, was too old to take to the field. He stood instead as the embattled country's legitimate political leader and last dynastic successor, taking his dead nephew's place, the very role it seemed he had been assigned forty years earlier. This assumption appears to have been borne out by Aedemon. He could have easily made himself king, but never, in fact, aspired to any royal pretensions, and continued throughout hostilities to serve only as Mauretania's top general, because Alexander Helios still sat on the throne, shaky as it was.

Although Cleopatra Selene had died at a relatively young age, longevity among members of the royal house was not uncommon. Juba's own grandfather Hiempsal died at the ripe old age of eighty-six. His great-grandfather Masinissa lived even longer, to ninety-two, and Juba himself continued to reign into his seventies. Septuagenarian monarchs like Alexander Helios were not all that rare in the ancient world.

However, written source materials for Rome's campaign against Mauretania are scant, leaving historians with assumptions based on few facts. Among these facts are Caligula's intentions regarding King Ptolemy. The accused monarch's death not only eliminated one of the conspirators, but also enabled the emperor to invade Africa Nova for a very cogent reason. Within three years after his ascension to the throne, Caligula had bankrupted the imperial exchequer through his lavish spending on luxurious entertainments and an insatiably opulent lifestyle.

Among his most extravagant whims was the construction of an immense, ceremonial bridge resting on a double line of ships extending for three miles over a section of the Bay of Naples. Accompanied by troops of infantry, cavalry, and the entire Praetorian Guard (which he had enlarged), Caligula drove across this impressive if otherwise useless engineering marvel just twice—coming and going in his golden chariot like the god he wanted everyone to believe he had become. For the occasion, he wore the original breastplate of Alexander the Great, brought all the way from the mausoleum at Alexandria. After sundown, the entire bay was aflame with thousands of torches. These costly extravaganzas were intended to show that, as the new Supreme Being, Little Boots dominated both the sea and the night. Even so, finding money to pay for such displays would tax even his divinity.

The spendthrift emperor could not fail to recognize the richest land in the empire. He knew that Mauretania's accumulated wealth was the result of more than sixty years of continuous prosperity. The revenues of the entire Mediterranean world had funneled into Caesarea since the industrious Juba II began his reforms back in 23 B.C. The dead king's gold would solve all of Caligula's financial problems with their serious political consequences. Foremost among these was payment of the Praetorian Guards. They were his personal protection against the senators, most of whom wished him dead. If he was unable to keep the Praetorians satisfied, they would lose enthusiasm for their imperial duties, and might even join his senatorial enemies. Painfully aware of these lethal possibilities brought about by his selfish draining of state finances, capture of Mauretania's treasury became Caligula's chief objective for the campaign. An invasion to seize the capital in a single stroke was immediately set in motion by powerful army forces, thereby preventing the enemy from escaping with the gold reserves.

A fleet of battle cruisers and military transports ferrying twenty-five thousand men in arms was under the able command of Marcus Licinius Crassus Frugi, who would afterward distinguish himself on British battlefields. His plan now was to brush aside any enemy vessels, then make a divided landing, simultaneously putting his legions ashore on either side of Caesarea and enveloping it in a single, lightning stroke. Mauretanian warships did indeed challenge their numerically superior opponents, offering a surprisingly spirited defense that disrupted the Roman timetable, but only slightly. The defenders suddenly broke off action, hastily withdrawing. The invasion had nonetheless been delayed enough to prevent Frugi's troops from disembarking at low tide.

A day late, the landings proceeded according to plan. The invaders encountered no opposition from the largely abandoned capital, but the treasury Caligula sent them to take had disappeared, along with the royal library. The Romans then broke into the mausoleum outside Caesarea to sack its treasures. But it too was empty. Even the solid gold sarcophagi containing the bodies of Queen Cleopatra Selene and King Juba were missing. From now on, the North African campaign became an increasingly desperate hunt for the missing Mauretanian exchequer.

Resistance grew more intense as the Roman forces penetrated deeper into the interior of the land. The enemy's hit-and-run sorties frayed nerves, slowed progress, and whittled away at supplies. By degrees of violence, enemy guerrilla tactics escalated into full-fledged battles for cities, towns, villages, and territories. The Mauretanian troops fought with unanticipated fanaticism and skill, the results of King Ptolemy's devotion to their previous years of training and outfitting. Most of the population had been roused by his execution. Now they were commanded by his right-hand man, Aedemon, who demonstrated a capacity for tenacity and adept maneuvering that any Roman field marshal could appreciate. Licinius had expected to encounter nothing more than disorganized tribal forays typical of rebellious provincials. Instead, he found himself fighting disciplined, well-equipped, highly motivated legions of infantry and cavalry in set battles worthy of a major opponent. Thirty years later, according to Tacitus (*History,* ii, 58), the Mauretanian army consisted of five *alae* and nineteen *cohorts,* roughly equivalent to modern divisions and corps, respectively—altogether, some fifteen thousand men. Given the size of the native male population, these numbers must have approximated the forces available to Aedemon.

Romans and Mauretanians were wrestling each other to death, and neither could quite get the upper hand. Their struggle seesawed across the whole country, demolishing everything from the Atlantic coast to the Atlas Mountains. Sections of ancient Volibulis, the second capital, went up in flames. Widespread burning swept the southwest area of the forum at Lixus, where the loss of life on both sides, including civilians, was prodigious. Fighting was so terrific at Tamuda that the entire city was burned to the ground, its smoldering remains and heaps of dead pillaged by local people made destitute by the carnage. The splendid kingdom that Juba II and his wife and son had built during more than six decades of unremitting toil was being shattered and laid to waste in flaming ruin. But the Mauretanians did not count the cost; nor did they surrender. While never quite able to beat Licinius decisively, they made him pay dearly in thousands of dead legionnaires, and left him nothing but ashes for conquest.

The war went on, month after month, but Licinius seemed closer to neither subduing the country nor capturing the enemy treasury. The emperor raged at the perceived incompetence of his commanding general, and at his presumed cowardice and disloyalty. An entire Roman army had failed to take over a miserable colonial province. Caligula sorely needed the Mauretanian gold—*now*. The rich treasure, however, was more elusive than ever.

In the end, he did not live long enough to lay hands on it. With the campaign still raging in Africa Nova, his career came to an ignominious end when the conspiracy he always dreaded struck him down on January 24, 41 A.D. Failure to obtain the Mauretanian treasury made his assassination inevitable, for without it he was unable to pay the Praetorian Guards, on whose personal protection his life depended. Indirectly, the Mauretanians, through their stiff resistance, had avenged the murder of their king.

Caligula was immediately succeeded by his uncle Tiberius Claudius Drusus Nero Germanicus (fig. 5.1). Unlike his egomaniacal predecessor, the new emperor was more interested in Roman justice than Roman circuses. He wanted to put an immediate end to the shameful war Caligula had instigated against the empire's best ally. But Claudius was pointedly instructed by the senate that the conquest of North Africa could not be called off under any circumstances. To do so would send the worst kind of signal to the empire's millions of colonial subjects, who would perceive his fairness as weakness and attempt their own independence movements. The imperial network that held together the whole civilized world would unravel in rebellions too numerous for even his legions to suppress. Other peoples considering revolt (and there

Fig. 5.1. Claudius, the successor to Caligula, was unable to stop the unjust war against Mauretania. (Capitoline Museum, Rome)

were many) must be discouraged by a Roman victory as ruthless as it was total. For the sake of the *Pax Romana,* an example had to be made of Mauretania.

There was another pressing incentive for its conquest. Caligula's profligacy had bankrupted the throne. By seizing King Ptolemy's treasury, a terrible economic crisis inherited by the new regime would be overcome. These political and financial imperatives persuaded Claudius that the admittedly unjust war needed to be won as speedily and completely as possible.

He found a new ally in Mauretania itself. While the majority of its people rallied to Aedemon's opposition forces, some sided with Licinius, actively joining his attempt to subdue the country. In a kingdom of such racial diversity, where each tribe had its own ethnic agenda and sense of identity, that at least one group went over to the invaders in the hope of fulfilling its own particular aspirations is hardly surprising. A Phoenician who Latinized his name to Marcus Valerius Severus, intending to curry favor with the Romans, headed auxiliary troops from half-ruined Volibulis against his own countrymen. He fought Ksar Pharaoun, a fellow citizen of the same city, who commanded a cavalry corps for Aedemon.

Although ethnic divisions undermined Mauretanian resistance, no one was ever in any doubt of the final outcome. Rome was approaching the zenith of her military might, and far more powerful states than King Ptolemy's kingdom had succumbed to her imperial demands. But the Mauretanians fought well, far better than anyone expected. Outnumbered and betrayed, they kept an entire Roman army at bay for seven months, an accomplishment few other enemies of the imperial eagle were able to achieve.

In late March 41 A.D., Marcus Licinius Crassus Frugi received the *ornamenta triumphalia,* the triumphal insignia, from the emperor for his final conquest of Mauretania. The formerly splendid kingdom was reduced once more to colonial status, divided by the river Malua into two provinces, Tingitana and Caesariensis, and administered by equestrian governors. Claudius had his conquest, plus the example he required to keep the rest of the world in proper awe of his authority. But he never found that other fruit of victory, the Mauretanian treasury. Its disappearance forms the rest of our story.

6

Escape or Die

Not lost, but gone forever.

SENECA

The preceding five chapters are documented history. Although investigators are sure to quibble over debatable dates, details, and interpretations, the events described actually took place. But with the military defeat and Roman occupation of Mauretania in early 41 A.D., the historical record breaks off, at least as far as the fate of its leading personalities is concerned. What remains to illuminate the unanswered questions left by those tragic events is primarily supposition based on inference. Surprisingly, however, an abundance of evidence, some of it circumstantial but much of it artifactual, can be organized like a grand mosaic, its fragments scattered by time and then reassembled into their original interrelationships. The image that thus emerges appears to be what subsequently took place.

Why did the Mauretanians bother to fight a war they knew could never be won? What became of their general, Aedemon? He did not fall in battle, and he was not paraded as a captive in the Roman triumph, for if he had fallen or been captured, contemporary historians would have surely reported it. Instead, he disappeared at the climax of the war. Much the same may be said for King Ptolemy's uncle and successor, whose reign lasted eight or nine months. The last surviving reference to Alexander Helios was to his arrival in Caesarea after the birth of Cleopatra Selene's son. But had he died anytime thereafter, the world surely would have been told. That he did not die before the war seems certain, while his participation in it as the legitimate monarch is likely, even though his name does not appear in the scant records of that time and place.

Most intriguingly, what happened to Rome's chief objective of the war, the fabulous Mauretanian treasury, to say nothing of Juba II's great library, his golden sarcophagus, and that of his wife? Again, had these treasures been simply looted or lost during the course of the fighting, their disappearance would likely have been cited by Roman historians. The Mauretanian exchequer was, after all, the war aim of both emperors.

To his people, Juba's library was the cultural equivalent of his royal treasury. His remains and those of his queen were the kingdom's most sacred heirlooms. All were continuously moved ahead of the enemy advance, beyond Caligula's grasp. Perhaps Mauretanian troops fought so ferociously against Roman forces, knowing full well they could never drive them back, in order to delay them, thereby providing enough time for the exchequer, other national treasures, and some privileged people to escape. Soldiers ordinarily fight to win, not to give their lives in a lost cause—unless they are convinced that by doing so they are sacrificing themselves according to the will of the gods. The suicide bombers who destroyed New York's World Trade Center in 2001 died in the name of Allah, just as less than sixty years earlier kamikaze pilots deliberately crashed their airplanes into American warships for the sake of Emperor Hirohito, Japan's living god. So, too, many Mauretanians had come to revere their dead sovereigns as gods, especially King Ptolemy, whose execution made resistance to Rome a sacred duty.

Moreover, Cleopatra Selene's notions of deified rulers carried over from her mother had undoubtedly taken root in North Africa through her promotion of the Isiac cult, to which many of Aedemon's troops belonged. The symbolic nature of her death during a lunar eclipse had gone far to make new converts to the mysteries she left behind and also deepened the convictions of their current adherents. Perhaps fighting and dying for their hallowed relics was victory enough for the Mauretanians. Or perhaps their chief motivation was the possibility of the escape of their cultural and economic heritage with survivors of the royal house, which would mean that the Mauretanian kingdom, if not the land itself, might be reborn somewhere else. But where?

"Members of the court knew that death or enslavement awaited them if caught," writes Professor Totten. "There was no hope for escape to the east or north, where Rome reigned. The route south along Africa's west coast was unpromising: tropical diseases, poor harbors, unnavigable rivers, and hostile locals."[1]

Here is how the story might have unfolded.

Everything north was enemy territory now, while the great waste of the Sahara Desert spread thousands of miles into the east. Alexander Helios, along with his court and military officers, was at the head of a mass migration of war refugees stampeding ever southward ahead of the Roman onslaught. It pursued them relentlessly down the foothills of the Atlas Mountains and across the length of present-day Morocco.

Even so, the furious resistance of Aedemon's dwindling armed forces continued to slow the invaders' progress. In fleeing south beyond the borders of Mauretania, the refugees had not been routed into unknown territories, but were beginning to make use of their Punic heritage. They knew that they had been preceded in this part of the world 465 years earlier by the Phoenician admiral Hanno, whose logbook was preserved in Juba's great library. A Greek copy survives to this day. On a mission to found new colonies in 425 B.C., Hanno passed the Senegal River, skirted the coasts of Sierra Leone and Ghana, then proceeded farther to the Cameroons before returning to Carthage. The information he and later Phoenician explorers brought back enabled the Mauretanians to find their way to West Africa and beyond.

Alexander Helios sent word to the chiefs of what is now the Tarfaya area of southern Morocco, then, as now, the home of black tribes on the West African coast. A meeting was arranged at which the services of men and materials were discussed. It was a difficult decision for the chiefs to make. To be sure, the abundance of gold being offered them was tempting. But if they incurred the displeasure of Rome, their own realms might be swallowed by the same kind of catastrophe that had just enveloped their white neighbors to the north. True, the Romans had not yet ventured this deeply into Africa, but what was there to stop them? And to what lengths were they willing to go for the capture of these fleeing Mauretanians? When Alexander Helios increased what he had originally offered to pay, however, the chiefs set aside their fears.

Work began at once on the construction of a new fleet. Under the supervision of Mauretanian maritime architects and engineers, native laborers cut lumber and laid the hulls of huge vessels the likes of which had never been seen before in West Africa. A number of Aedemon's warships already rode at anchor offshore, but there were not enough of them to accommodate tens of thousands of refugees. These comprised the entire Mauretanian court; numerous aristocrats with money enough to buy their way out of the destruction of their country; many army officers in command of elite troops chosen to protect their privileged charges; and a company of professional seafarers, experienced navigators and captains.

Deckhands were in particularly short supply, however, so West Africans were asked to help make the long, hazardous journey ahead. They were joined by affluent Jews who had once played influential roles in Mauretanian government and commerce. An even smaller minority of early Christians won passage because they had been able to make some converts among members of court who were convinced the traditional gods had deserted them with the fall of their country. Spread throughout this polyglot company were adherents of several mystery religions, from those who followed the old Carthaginian worship of Tanit to the arcane Gnostics.

Regardless of their spiritual beliefs, all of the refugees were under a sentence of death. If the Romans caught up with them, every man would be crucified, their women and children sold into lifelong slavery. The circumstances of defeat had not only obliterated their kingdom; it had also reduced them to the status of first-century "boat people" who were forced to risk their lives on dangerous seas rather than certainly lose them by remaining behind on land. All too conscious of the fate looming in the north, the Mauretanians and their West African coworkers labored ferociously to finish the new ships. There was another urgent consideration: the weather. Spring was waning, and if the fleet was not ready by midsummer, it would not be able to sail until the following year, which would be too late for them to escape the doom awaiting them. November marked the beginning of storm season, and the refugees must arrive at their distant destination before then. Otherwise, the prospect of surviving the prolonged voyage in makeshift vessels was bleak. Urgency gave impetus to their efforts, and the hastily completed ships were boarded a few weeks after the summer solstice.

The priests and holy men representing a half dozen or more different faiths pleaded with their own versions of God for his blessing on the desperate adventure. The great vessels, heavily laden with passengers and supplies, lurched out over that vast, rolling carpet of the heaving unknown that Roman cynicism disparaged as "the Pasture of Fools." Crowds of West Africans gathered on shore with their uncertain monarch to watch the strange craft carrying their strange passengers and crew farther west with every foaming plunge. At the same moment, those on deck watched with the heavy hearts of displaced persons as the continent, if not in every case the country of their birth, fell gradually but forever away to stern.

But they had not thrown themselves aimlessly into the Atlantic Ocean just to escape their pursuers. They may have known exactly where they were going. The seafaring heritage of Mauretania was already centuries old when King Juba II sent out expeditions to the Canary Islands, even as far as Madeira, in search of murex mollusks for his lucrative purple-dye industry. Henceforward, the classical world referred to these distant shores as the Purpurariae, or Purple Isles.[2]

Juba's great library contained his Punic legacy of sailing instructions to distant, prosperous locations unknown to the outside world. These were jealously guarded by Carthaginian merchants as treasured commercial secrets. Charts drawn from an abundance of such privileged information enabled the captains of the refugee fleet to find their way to what the Roman cartographers vaguely knew as the Epeiros Occidentalis, conjectured continents that most likely lay in the far west, beyond the horizon that only suicidal fools dared to surpass.

Actually, the existence of both continents had been known to the Romans for at least two hundred years, even if they were unaware of how to find them. A globe of the world created by the Greek geographer Crates of Mallos and preserved for the Romans in the writings of Strabo identified North America as Peroikoi and South America as Antipodea. As Norman Totten, professor of history at Bentley College, in Waltham, Massachusetts, and professor of anthropology at Bridgewater State College, in Bridgewater, Massachusetts, explains, "Diodorus of Sicily, noted historian, wrote in 21 B.C. that Carthaginians had known about lands many days' sail west of North Africa. According to Diodorus, the islands had mountains and plains, navigable rivers, and lush vegetation. These references seemingly point to America. Caribbean islands fit the description, but other islands in the Atlantic do not. Diodorus wrote during the early part of Juba II's reign that Carthaginians kept this knowledge secret in order to have a place of refuge, if needed."[3]

To resume our story, while a July or August passage across the tranquil central Atlantic may have been tedious, it was, outside of an isolated storm or two, mercifully uneventful. The point from which the expedition left Africa would have thrust it directly into the Canary Island current flowing westward straight across the Atlantic to the opposite side of the ocean. (This was the current that carried Christopher Columbus's *Nina, Pinta,* and *Santa Maria* to the New World, and later brought the primitive reed boat of maverick archaeologist Dr. Thor Heyerdahl safely to Barbados from the same Moroccan shores King Juba's people may have left nearly two thousand years earlier.)

As their ships approached the Gulf of Mexico, however, hurricane season arrived. Some vessels must have been sunk and the lives of those aboard lost. Others were probably separated from the fleet and scattered to the coasts of Florida, Mexico, and Panama. These areas had been specifically warned against in the Phoenician records as landfalls to be avoided at all costs, due to the savagery of their inhabitants. Florida's Arawak Indians were engaged in constant cannibalistic warfare with tribes of Caribbs. The Mayas of Yucatán, while more civilized, nonetheless delighted in human sacrifice and ritual torture. Panama was a perilous wilderness of still more violent natives and deadly disease.[4]

In spite of these dire admonitions, it seems that the lost Mauretanians landed on the eastern shores of Central America, moving on from Panama to leave their mark on other parts of South America, most notably in what are now Ecuador and northern Peru. These storm-blown arrivals represented the smaller contingent of Mauretanian refugees, and they likely disappeared after only a few generations through the inevitable combined attrition of warfare, intermarriage, and contagion.

The main body of immigrant vessels, however, entered the mouth of the Mississippi River, sailing northward into the North American continent. Such a passage would not be possible today. With the construction of wing dams in the late nineteenth century, but especially after the installation of locks during the 1930s, the southward-moving current of the Father of Waters increased to its present seven-mile-per-hour flow. Even the antebellum steamboats described by Mark Twain would have lacked sufficient power to navigate the river in its present condition. Little more than one hundred years ago, however, the river was virtually stagnant, its current negligible save in early spring, when meltwater from previous winter ice and snow raised its level and speed. At other times, it allowed Indian canoes to be paddled with ease in all directions, something eighteenth-century visitors like Jolliet and Marquette learned during their explorations of the Mississippi Valley.[5] The Mauretanians found the same navigable circumstances, and, if they arrived from mid- to late summer, strong southerly winds would have speeded them on their way northward.

They undoubtedly pulled into shore from time to time for repairs and other necessities. At such landing intervals, foraging parties hunted game and brought back edible fruits. It was then that the vital presence of so many armed troops on the expedition became apparent. Various indigenous groups encountered along the way made mostly furtive appearances, keeping their distance from the awesome strangers. Often there were ambushes, surprise assaults. But Aedemon's professional soldiers, veterans of the recent life-and-death struggle against faraway Rome and earlier anti-guerrilla warfare campaigns against Tacfarnias, were never caught unawares, and beat off every attempted sortie, no matter what the numerical odds.

Not all Indian encounters resulted in violence, however. Some tribes showed more inclination to barter than fight, and friendly relations with several river peoples were established. Among these, the West Africans attracted special notice for their skin color, in the same way unusual or unfamiliar personal traits were regarded as the sure signs of *wanaka,* or spiritual power, as it was known throughout Native American tradition. Many blacks jumped ship, preferring to settle where they were clearly wanted rather than entrusting themselves to the less certain destiny of the white foreigners they served as deckhands. Moreover, Louisiana's temperatures approximated the climate conditions they'd left behind in West Africa, while the climes of the North, where the Mauretanian expedition was headed, were unknown.

Actual evidence of their decision to abandon the other refugees survives among a numerically insignificant population of native blacks known today as the Washitaw. They were described as an indigenous people by the renowned explorers Lewis and Clark in 1797. Four years later, the Washitaw, as a sovereign nation, were distinguished from the region's Indian tribes and were officially exempted from the Louisiana Purchase. In 1995, the Economic and Social Council of the United Nations officially recognized the Washitaw blacks' indigenous, pre-columbian identity.

Indeed, a present Washitaw leader, Verdiacee Tiari (see fig. 6.1), had been told as a child that her ancestors were not brought over to America as slaves, but instead arrived as free men and women many centuries earlier. This oral tradition had been passed down from her grandmother, and from her grandmother before her, and so forth over the course of numerous generations.

While some of their West African deckhands were jumping ship, the Mauretanians continued to push toward the North in a prolonged expedition that took almost as long as their transatlantic crossing. When they at last reached the Mississippi River's confluence with the Ohio River, autumn was spreading its swirling colors. The newcomers were not the first to visit this spectacularly raw wilderness. They had been preceded centuries before by Phoenician pathfinders who came in search of the rare goods that once made their Carthaginian capital the richest city in the world. Their detailed exploratory knowledge of the "opposite continent," actually cited as long ago as the fourth century B.C. by the Greek philosopher Plato in his dialogue the *Kritias*, was classified information, however. Preserved as a secret legacy by Juba II, it had since passed on as a new birthright for his uprooted subjects.

Fig. 6.1. Verdiacee Tiari, leader of the Washitaw Nation, claims her African heritage in Louisiana predates modern European arrivals by many centuries.

Steering east into the Ohio River, the refugee fleet followed its course in what would millennia later become southern Illinois, traveling northward into the Wabash River, with the future state of Indiana to starboard on the east. The ships then turned once more, west this time, into a smaller Illinois river christened the Embarras (pronounced Ahm-bahr-ahhs) by French traders some seventeen centuries after the Mauretanians' time. Just two more days on the water, and the long journey from Africa was over. It had taken the immigrants about five months to complete their flight from the vengeance of Rome, which would never find them here.

The Mauretanians dropped anchor at a very definite place pinpointed on the old Phoenician chart and carefully chosen for the high purpose that had originally prompted their arduous voyage beyond escape and survival. Not far from the south bank of the river, they began digging almost at once, excavating and modifying the landscape. Military engineers ordered their troops to look lively. A chill was already in the air, presaging winter. If the ground froze, the task for which they had traveled so many thousands of miles would not be finished in time. They leaned on their entrenching tools, gradually converting a natural cave system into lengthy corridors interconnected with numerous separate chambers. This fundamental job complete, artisans and craftsmen entered the underground spaces, installing oil lamps and wall decorations, mostly stone friezes depicting scenes from the rituals performed by initiates in the holy mysteries of Isis.

When work on the subterranean project came to a halt, the Mauretanians gathered after nightfall in their finest apparel for a sacred ceremony. Under the spectral glare of flickering torches, they sang hymns and chanted prayers, while the men with their commanders stood silently at attention, rank upon rank, fully armed and in dress uniforms. A volley of trumpet calls from the nearby ships silenced the singing and chanting, simultaneously snapping every officer and soldier to eyes right. The sound of a dirge intoned by the deep male voices of religious functionaries echoed from the riverbank as a cortege approached with slow dignity. When it reached the crowd of mourners, growing louder in its lamentations, the people saw their last monarch, the elderly Alexander Helios, borne in the lead on a raised throne by four holy men. He was dressed in the white linen robes of a high priest in the cult of the goddess.

In one mass gesture, the soldiers drew up at attention, unsheathed their swords, and saluted in the Roman manner, with the naked blades at arm's length. Behind their sovereign, the solid gold sarcophagus of his twin sister, Cleopatra Selene, followed on its own funeral caisson. Next came another gold casket, almost identical in its ornate craftsmanship save for the name inscribed in Egyptian hieroglyphs on the lid: Juba II. King and queen were interred in the newly completed tomb amid secret rites only the priests and their immediate assistants were allowed to observe. Then Juba's encyclopedic library was buried with him. Finally, large, ponderous coffers were carried, one after the other, into the bowels of the underground chambers.

For the next 1,941 years, the royal treasury of Mauretania would lie undisturbed beneath the soil of Illinois.

7

Discovery in Southern Illinois

What merited silence has been told, and what deserved telling has been passed over in silence.
THEODOR MOMMSEN, THE PROVINCES OF THE
ROMAN EMPIRE FROM CAESAR TO DIOCLETIAN

The countryside of southern Illinois is remote, a practically forgotten cultural backwater somewhere between St. Louis, at the western border with Missouri, and the state university in Carbondale, forty-five miles north of Kentucky. The inhabitants would have it no other way. Their numbers are small and disparate. Although income and educational levels are below national or even state averages, people are hardworking, Bible-conscious, gun-owning patriots residing mostly on old, isolated farms or in charming, unprosperous little towns. Folks are friendly toward as well as wary of strangers. They prefer their largely anonymous, unvisited status. Attitudes can be provincial and territorial, while speech patterns echo those from below the Mason-Dixon Line. Among landowners there is a highly developed sense of protective sovereignty regarding the properties they own and on which they grow their crops, which are mostly beans and corn.

Southern Illinois has always been a refuge for rugged individuals. Local history tells of frontieresque lawlessness and even gang wars with criminal interlopers from Chicago, such as Al Capone and Tony Accardo (during the Roaring Twenties and 1960s, respectively). Much earlier the infamous Harpe brothers, Micajah and Wiley, murdered some fifty victims at Cave-In-Rock on the Ohio River, before they were beheaded in 1799.

On the eastern side of the state, directly across from St. Louis, Richland is the second nearest county to the Indiana border. Beyond its sparsely inhabited hills and ravines, squares of brown-green farmland spread toward the horizon like pieces in an agricultural puzzle. In the extreme northeast corner of Richland County, an elbow of the Embarras River bends into Illinois from its bigger sister, the Wabash. Locals have for generations enjoyed exploring or picnicking in the numerous caves that honeycomb the area. One infrequently visited site, certainly unknown outside its immediate vicinity, was hardly more than a hole in the ground. But the opening, about ten feet wide and eight feet from floor to ceiling, was large enough for visitors to pass through, stooped, into a kind of natural corridor that ran about fifteen feet into the side of a hill, perhaps three quarters of a mile from the south bank of the Embarras.

At the far end of this seemingly insignificant cave was a small chamber. Whether it was natural or man-made was difficult to determine. Its walls were decorated with what visitors assumed were "Indian signs"— apparently old carvings of bizarre animals, inscrutable glyphs, and strangely costumed men—rendered in faded, primitive stick form. Obviously, the cave had been used by Kickapoo or Shawnee tribes who inhabited the Richland County region into the early 1800s. No one gave the place a second thought until 1982. Certainly, professional archaeologists, if they even knew it existed, never declared the site off-limits to public entry, nor forbade anyone from doing what he pleased there.

But on April 2, 1982, a forty-seven-year-old "caver," outfitted with flashlight, pick hammer, and knapsack, entered its dark recesses. He had come from his home in Olney, a small town about fifteen miles away. Born in West Virginia, Russell E. Burrows had moved to southern Illinois after a stint in the U.S. Army during the Korean War, and it was there that he developed an interest in local history and began amassing a collection of everyday objects from the past.

Fig. 7.1. Russell E. Burrows stands atop a hill in Richland County, Illinois, in March 1998. Pointing downward, he proclaims, "This is the Cave!" Photograph by Wayne May

Over time, he found ox shoes, square nails, iron pots, lanterns, and other nineteenth- or early-twentieth-century artifacts for his growing collection. A woodworker by vocation, he could appreciate these handmade items of yesteryear. Perhaps, he thought, something of the kind might be found in the curious little cave he had heard tell of in the northeast section of the county. Upon visiting it and finding the area outside its entrance deserted, as it usually was, he paused momentarily to scan an uncertain sky, as muted thunder boomed ominously in the distance.

He found the interior as described—a small, unimpressive natural enclosure resembling others he knew. Proceeding to its apparent termination, Burrows stepped into the close confines of the chamber. He wondered if perhaps the space was artificial, though why anyone would go to the bother of carving it out of the earth made no sense to him. He recalled, however, that it was not unusual for Indian creations to elude the understanding of modern white men. The glare of his flashlight passed over a series of crude drawings adorning the walls here and there, which might make colorful additions to his rather lackluster collection of common pioneer knickknacks. Clearly, there were no nineteenth-century hobnails lying about.

With the first taps of his hammer against the wall, however, he noticed something strange. The impact of his tool against stone did not make quite the solid sound he expected. Instead it produced a lighter reverberation, as though a hollow space lay on the other side. Curious to learn if a cavity did indeed lie just beyond, he swung his pick against the face of the wall. As he labored with a will, he was encouraged by what seemed like the echoes of his hammer blows coming from some place deeper in the hill.

The work was difficult, but Burrows was a strong man, and after perhaps fifteen minutes of sweat and effort, the stones in the wall began to give. Suddenly, they tumbled heavily away, thudding to the ground and disclosing another small chamber, this one unquestionably man-made. It was the opening to a flight of stone steps leading down into the earth. He played his flashlight over them, then carefully followed its illumination into the otherwise impenetrable darkness. The flight of stairs was steep, and he descended cautiously, sideways, eventually reaching bottom. He estimated it was about thirty feet from the entrance above. A long, dead straight corridor disappeared into the darkness before him. Burrows's bright flashlight lit up its still, dank interior as he carefully entered.

The tunnel, perfectly hewn, was hung with very old-fashioned oil lamps at regular intervals. They looked like something out of a movie he might have seen once about ancient Rome. He proceeded cautiously. The atmosphere was heavy with mystery, and he recalled that snakes, especially deadly copperheads, were known to favor such subterranean environments. But he encountered no serpents; the muted sound of his own footfalls in the almost stifling confines was all he heard. The tunnel went on and on, leading past dozens of dead oil lamps on either wall. Turning the beam of his flashlight at the low ceiling, he saw that it was covered with black smudges, the residue, apparently, of innumerable torches that were once carried this way, though how long ago he could not guess.

After Burrows had walked about five hundred feet, the corridor, seeming to come to an abrupt end, actually made a sharp right turn. His flashlight illuminated another great length running straight ahead beyond the white reach of its flickering bulb.

He proceeded a few paces until a low open portal, minus a door, appeared unexpectedly on his left. Ducking down under its low lintel, he entered a small chamber, then almost at once staggered backward in surprise. Gleaming in the harsh beam of his flashlight stood the five-foot-tall statue of a man, wrought in solid gold. Nor was this just the representation of any man. Its beneficent pose and holes in the wrists of the outstretched arms clearly identified the figure. A few feet behind the statue, to its left, was a raised platform perhaps three feet high. On it had been laid a full-size sarcophagus, likewise executed in gold. Recovering from the shock of his discovery, Burrows breathlessly admired the spectacular craftsmanship of both objects, but refrained from touching them. He could hardly believe what he saw. After he left this chamber, he found and entered several more rooms in quick succession.

Across the floor of one were stacked edged weapons—a metal sword with shield and battle-ax together with a set of bronze spears ranging in length from three to six feet. There was copper or bronze armor—breastplates and greaves, even helmets. Nearby stood stone statuettes of noble-looking men and women dressed in strange garb suggestive of the ancient Nile Valley or Carthage. Stone and clay-fired jars or urns, some of them half as tall as a man, were positioned in two corners at the far end of the room. A number had long ago fallen over and broken open to reveal their contents—leather or hide scrolls covered with an inscrutable written language. Scattered among these jars were smaller oil lamps like those attached to the walls of the corridor, and paint pots.

A recessed shelf cut into the stone cave wall and, supporting the sculpted images of Egyptian-like deities, ran around the whole interior of the enclosure. Against one wall were piles of perhaps one hundred flat black stones, each one engraved with a human profile and unreadable inscription. The faces portrayed a bewildering variety of men and women—though mostly men, depicted as soldiers in Roman-style helmets or as priests in robes. They showed European or Semitic facial features, but wore the togas and uniforms of civilizations long since passed into history.

Stepping into an adjacent chamber of similar dimensions, Burrows noticed a vault cut into the rock face of the cave. It shone in the glare of his flashlight with numerous piles of gold coins—what would later prove to be more than a ton. This same vault contained a quart-sized stone bowl filled with uncut diamonds. Nearly faint with these discoveries, he played the flashlight in his trembling hand over the far wall of the chamber, and saw at once that it opened to another. This room was much larger, about twenty by twenty-five feet, at the center of which lay a large stone sarcophagus. Inside was a gold coffin of superb workmanship. As in the other chambers, enormous piles of black stones emblazoned with lengthy, peculiar inscriptions, strange symbols, and the images of both human beings and animals filled the crypt. The people portrayed were an impossible mix of apparent Romans, Phoenicians, Hebrews, Christians, American Indians, and even black Africans. Some of the animals depicted on the stones, such as lions, elephants, and camels, were not native to America, at least not since the last Ice Age twelve thousand years ago. Yet here they were, represented in all their incongruity.

The unreality of this subterranean site was making him dizzy. The atmosphere was stiflingly close with some nameless presence. He needed fresh air, to get back into the upper world. Returning to the chamber where he'd seen the gold, he availed himself of as much booty as he could carry, then hurried at all speed, his bulging knapsack and sagging pockets clinking with gold coins and several dozen diamonds.

In moments he was scrambling through the broken wall, back into the little chamber at the back of the cave. Burrows was elated by his incredible good fortune. It was the find of a lifetime. Clearly, whatever this place was, he thought, its importance and wealth were too great to leave unguarded. He tried to reassemble the old wall he'd hammered apart, but anyone who happened to see the repositioned stones would know they had been recently dislodged. The cave, though rarely visited, was now especially vulnerable to inquisitive people like himself. Others might discover the break-in and loot the rest of the treasures. Emerging into the open air, he was relieved to find himself still alone.

Because he could not hope to restore the collapsed wall to the condition in which he'd found it, Burrows concealed the cave opening itself. He hadn't forgotten what he'd learned in the Army during the Korean War; he dragged shrubs and tree limbs over the gaping hole to camouflage its appearance, then realigned large stones to alter the opening and immediate surroundings. Within an hour, the cave was so thoroughly disguised that no one who was not intimately familiar with its vicinity would ever locate the entrance. Satisfied that his find was safely hidden under the subtly altered environment, he returned to his pickup truck two hundred feet away. Afternoon declined toward evening. Deep shadows were already filling gullies and ravines. They obscured the location even more effectively than his concealment of foliage and rocks.

The fabulous find was his by right of discovery, regardless of who happened presently to own the property on which it was found. And it would remain his as long as he preserved the secrecy of its whereabouts. No matter who may someday try to claim it, he mused to himself as he trudged through the lengthening twilight toward his home, the site would hereafter and forever be known as Burrows Cave.

Nineteen years later, Russell Burrows publicly presented a detailed description of the events of April 2 before an international archaeology conference at the Vienna Art Center in Austria, explaining the depth of the cave to the terminal chambers (535 feet), the angle of descent of the passageway (six degrees), the nature and location of the items he discovered, and the lamps affixed to the walls.

Remarkably, the cavern's dimensions and features resemble those of Kubr-er-Roumia, King Juba II's mausoleum, from which his mummified body and treasure trove were removed ahead of the Roman invasion of 40 A.D. The first professional investigators of his tomb "found themselves in a long gallery about eight feet high and six and a half feet broad. There were niches along the walls which seemed as if they had been made to hold lamps," according to A. MacCallum Scott.[1]

Burrows offered that the statues he discovered in the first chamber were "arranged in a semi-circle" and appeared "on the order of Egyptian figures, the left foot forward and the left arm forward," with an estimated height of eight or more feet and an estimated weight, because they were made of an apparently dense and heavy black material, of four to six tons each.[2]

Burrows testified to the conference-goers that thirteen doorways were cut into the walls of the cave:

> These doorways are closed by cut and well-fitted blocks of stone, the seams of which are sealed with a pitch or beeswax. I removed one of the blocks, and was amazed to discover that the sealed doorways were the entrance into a burial crypt, which was about twelve feet square, with a stone bier in the center. In this crypt, I found the skeleton of a male; this was determined by the pelvic bone.[3]

And he went on to describe the copper, gold, jewels, ax, and shield discovered within this tomb, as well as the scrolls with strange writing, which, he assured the archaeologists, he did not touch. Burrows then outlined his discoveries in the two other burial vaults, one holding the remains of a woman and two children:

> In the area of the heart of the woman was embedded, through the rib, a golden blade large enough to have penetrated the heart. It appeared to me that . . . the blade . . . had become "locked" in place by bone, so that when the effort to remove it was made, it was pulled loose from its shaft, and was left in place. The children each had a large hole in the forehead. Lying on the bier with the remains were two ax heads made of pure white marble. One of these axes fit the holes in the children's heads perfectly.
>
> Further back and in a lower level of the cave is another burial crypt, which is much larger and different in that there is a sarcophagus in the center which has a stone lid closing it. Inside is to be found a fine golden coffin much like those seen in Egyptian burials. Inside the coffin is . . . what appears to be [a] mummy. I cannot state for cer-

tain that that is the case, because I did not disturb the decaying cloth around the body.[4]

Burrows also found in this deepest chamber a round, wheel-like device made of rock, which seemed to be the mechanism used to close the crypt, perhaps by dropping down into a trough and rolling, in pulley fashion, after the cord holding it was cut. Along with the sarcophagus there were statues of a figure resembling the Egyptian god Amen-Ra. He stressed a number of times in the presentation that he disturbed none of his findings.[5]

Listening to his matter-of-fact presentation delivered in a steady West Virginia drawl, the continental scientists assembled in Vienna's meeting hall were stunned. Such a tale was utterly beyond belief. But there was more than narrative to Burrows's fantastic story. Much more.

8
Gold, Archaeological and Otherwise

I'm sending along $150,000 of gold.
BURROWS CAVE'S "LANDOWNER"

Russell Burrows's first impulse on beholding the cave's stacks of gold and bowl of diamonds had been a naturally human one: to pocket as much treasure as possible now and return for the rest later. For some years, a close friend, Wayne May, publisher of *Ancient American* magazine, eventually became so exasperated with broken promises to be shown the cave that he told *The Barnes Review,* a nationally distributed history magazine, that "Burrows melted the gold down and sold it. He looted the site."[1]

Russell claims not to have yielded to greed, however. He knew little about archaeology, but had recently read newspaper reports of a new Illinois law that imposed harsh penalties, including imprisonment, on anyone tampering with or confiscating prehistoric valuables. In addition to the gold and diamonds, several of the black tablets incised with peculiar written language and the profiles of what seemed to be ancient strangers had also been removed. Had he already broken this law by taking a few of the inscribed stones? Should he report his discovery? But to whom? The police? If the site's location became generally known, it would certainly be pillaged by people from miles around, and he might be blamed for it.

Fig. 8.1. A fraction of the golden hoard from southern Illinois's elusive cave. Photograph republished with permission, *Ancient American* 3, no. 16 (January/February 1997).

Almost immediately after making his discovery, Burrows sought out an old friend, Thelma McClain, an amateur archaeologist with her own large collection of Indian arrowheads and pottery shards. Each specimen had been gathered by hand from Illinois prairies over the previous twenty-five years, and some of them were for sale at her small antique-curio shop in Olney. In sworn secrecy he told her all the details of his discovery, except its precise whereabouts. Although he was sure she would never betray his confidence, he thought she might inadvertently divulge the location, thereby endangering it. She listened attentively to his story but found it difficult to believe—until he showed her the stones he removed from the cave. They were like nothing in her collection of Illinois Indian relics, but there was something familiar about a line of script on one of the stones. Referring to a book on comparative languages in the ancient world, she located a line of written characters identical to those on one of the cave's inscribed tablets. The language was West Semitic, spoken by Carthaginians more than two thousand years ago. On the opposite side of the stone was the illustration of an unusual vessel with a deep hull and prow fashioned to resemble a bird's head.

In *Ships and Seamanship of the Ancient World* by the foremost authority on the subject, Lionel Casson, McClain found the page she'd had in mind and pointed to the drawing of a strange sailing vessel that looked almost the same as the ship carved in the stone (fig. 8.2). The vessel was Phoenician, just like the extinct language written on the tablet.

Burrows certainly didn't know enough about Phoenician or Carthaginian history and archaeology to have faked the pictures and

Fig. 8.2. The stone at left, from the southern Illinois cave, depicts a sailing vessel that is markedly similar to the representation at right of Phoenician warships from Carthage (circa 170 B.C.). Note that the fish and oversized heads in square portholes are stylistically identical.

writing on the stones. After he told McClain that there were perhaps thousands of stones in the cave like those he had shown her, they tried to decide what should be done next. Should they go to the state capitol in Springfield?

McClain could recall stories from fellow collectors of unusual artifacts about the disappearance of objects that they'd lent to scientific authorities for analysis. While such rumors were probably not true, she nevertheless decided with Burrows to take the cave's artifacts to Jack Ward first.

Jack Ward (fig. 8.3) ran a stone-crushing company in Vincennes, Indiana, where he had been nominated director of the Sonotabac Prehistoric Indian Mound Museum. Although not a university-trained archaeologist, he participated as a volunteer assistant in professional digs as early as 1931, when he was in on the discovery of a major site in Wickliffe, Kentucky, just across the Ohio River from Illinois. He helped excavate a Hopewellian village, complete with dozens of human remains and abundant copper awls, fishhooks, and spearheads. Hopewell, from the farm after which the first artifacts of their kind were found, refers to a sophisticated culture that flourished in the Midwest (mostly in Wisconsin, Illinois, Indiana, Kentucky, Michigan, and Ohio) from about 200 B.C. to 400 A.D.

In 1933, Ward assisted the Illinois State Archaeologist in excavating another prehistoric location at East Carondolet. Five years later, he helped save Monks Mound, at Cahokia, across the Mississippi River from St. Louis, Missouri. For two hundred years, beginning around 900 A.D., Cahokia was the capital of a far-flung civilization known as the

Fig. 8.3. Jack Ward appears on the cover of *Ancient American* magazine surrounded by gold objects removed from the southern Illinois cave.

Upper Mississippian culture, with trade connections as far away as the Rocky Mountains and the eastern seaboard. Monks Mound (named after some Trappist monks who resided briefly at its summit in the late eighteenth century) is the largest earthen structure north of the Rio Grande, with a fourteen-square-acre base—larger than that of Egypt's Great Pyramid. In 1938 it was threatened with destruction by development and Ward had been asked by Senator Joe McGlyn to lobby the State of Illinois on behalf of Cahokia's preservation. Today it is recognized as one of the premiere archaeological sites in the United States.

Describing Jack Ward in *Ancient American* magazine, author Horatio Rybnikar wrote:

> A pillar of the Vincennes community, which recognized him as Citizen of the Year on several occasions, he was on personal terms with Presidents Eisenhower, Kennedy, Johnson, and Reagan. In 1983, he was presented with a special award signed by Reagan for outstanding public service. Jack Ward was familiar with senators, congressmen, and judges in his home state. He was regularly consulted about important government projects and assisted in items of Indiana legislation. He was the leading figure and champion of the Wabash River Transportation Project which increased maritime navigation along the Wabash River and its important tributaries. He was selected to oversee, document, and print all data relating to the Wabash River Aquifer that flows along the New Madrid Fault below the Wabash. He researched and recorded for the states of Illinois and Indiana the final government drafts concerning this strange and magnificent underground river with a flow of over 125 trillion gallons per day beneath the Wabash. His published thesis is still used by the State of Indiana as the authoritative manuscript on a subject affecting millions of residents in the tri-state area.[2]

By the early 1980s, Ward's reputation as an experienced, respected antiquarian won him election as head of the Sonotabac Museum. His first act as director was to donate most of his extensive Indian artifact collection, which had been accumulated over the previous fifty years. But nothing among his vast array of flint arrowheads and broken pottery prepared him for the inscribed black stones Thelma McClain showed him.

"With the utmost skepticism," he later recalled, "I looked at almost twenty or thirty of the items." Burrows had by then returned to their subterranean source several times for additional samples. "I spent two hours questioning him, and he explained his reluctance to disclose the cave's location, which to me was a very valid reason."[3]

Ward was still doubtful until he was shown one stone bearing a lengthy inscription he recognized as Phoenician. Soon, he was convinced that "these artifacts are of genuine prehistoric origin." For the next several years he "continued to acquire artifacts from the cave from Mr. Burrows for study to determine just what these things are. I have been assisted in this acquisition by Mr. Norman Cullen. These purchases from Mr. Burrows assist him in defraying his expenses for exploration."[4] Norman Cullin was Ward's closest friend, an Indiana contractor, and, like him, an amateur archaeologist. Both men eventually bought dozens of inscribed and illustrated stones from Russell Burrows, so many, in fact, that the three men formed a joint association to preserve and promote the cave finds. They called it the Artifact Recovery Exchange (A.R.E.). Ward was its administrator and Cullen his assistant, with Burrows the official explorer. Many of his stones were stored at the Sonotabac Museum. By April 1986, more than two thousand ("no two are alike"[5]) were brought there for display.

Local reaction was more mixed than enthusiastic. While many visitors expressed wonder at such items, they doubted their authenticity, especially because Burrows steadfastly refused to show anyone the cave. But Ward explained:

> [T]he owner [of the land on which the site is located], who lives in a distant state, was adamant that his identity not be revealed, primarily because he didn't want hordes of curious people invading his property. I have gained the landowner's signed permission to take, at my discretion, any qualified person onto his property and into the cave. During our negotiations, I learned that the landowner has known me since 1932, and even knew my father. However, I am determined to

respect the wishes of the landowner and also respect his reasons for maintaining anonymity.

In one of his letters to me, the landowner tells of his going into the cave with Mr. Burrows, and how he himself probed into the silt floor with his own hands to recover a stone sculpted face of Nubian characteristic, and personally sent it to me. He confirmed that he saw with his own eyes the five life-size statues carved out of diorite in the farthest extreme chamber, which is two hundred feet in from the entrance to the cave. Even though he has seen that the cave contains ancient artifacts, he is not interested in obtaining anything monetarily for any of the artifacts, and maintains that he has given Mr. Burrows free permission to retrieve and do what he wishes with anything he finds in the cave.[6]

According to Ward, only four people in addition to Burrows visited the interior of the cave, including himself. These were his colleague Norman Cullen, Sam Eyer (Burrows's brother-in-law, who helped him carry out some stones), and the unnamed landowner.

What about the gold and diamonds? Burrows insists the latter are still in the cave, mostly untouched since he first saw them in 1982. The gold—all one ton of it—was removed and taken to the Sonotabac Museum. There it was photographed, weighed, and studied. Every piece was then laboriously returned to its original position in the second chamber.

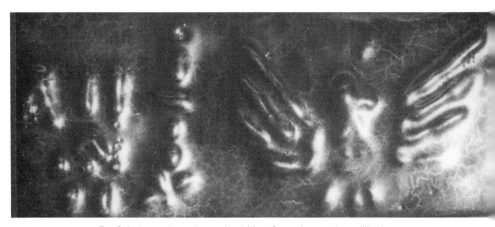

Fig. 8.4. An eagle-embossed gold bar from the southern Illinois cave

Fig. 8.5. A gold piece from the southern Illinois cave with an inscription in ancient Numidian, the mother tongue of King Juba. Photograph by Beverley Moseley

Most people who have heard this version of events find it very difficult to believe. One of them is Harry Hubbard, an *Ancient American* writer who investigated the story. "The records I have [of gold removed from Burrows Cave] only cover the summer of 1987 to fall of 1989," he stated.

> During that period, $6,995,308.12 worth of gold passed through the office at A.R.E. . . . It appears that a major portion of the first dump was melted down and sold to the U.S. Mint through Fort Knox. The cash was then transferred by the company in a Swiss bank account under their [the A.R.E.] partnership agreement. Burrows was afraid the Internal Revenue Service would become suspicious if they spent it or stashed it here. I can't find in my records where the rest of the gold ($3 million worth) went. I believe Burrows still has it stashed away, probably in Olney (the town nearest the cave) somewhere . . . Ward had the gold tested for purity, and it came back with a certification of over 98 percent pure. . . . The facts bear me out. If I'm wrong, I'll admit it.[7]

In 1997, Hubbard reproduced copies of endorsed correspondence "describing vast amounts of money (in excess of $160,000) collected in less than a week from the sale of gold, silver, and artifacts" found in the cave.[8] He included "a report on the events of August 25, 1987," submitted to the A.R.E., in which Burrows described a visit to the site with the landowner of the property on which it was located:

> We entered the cave and went to the room with the statues. He said that he wanted me to recover as much of the gold as possible as soon as possible. When asked why, he said that we have some problems coming and we should get as much as possible beforehand. I asked him what kind of problems and he said he didn't know, but that this time of the year we usually have problems. He and I carried out a total of just over five hundred ounces.[9]

A little more than three months later, Ward and Cullen received a letter dated December 6 from none other than the previously unnamed landowner, George Neff.

> Russell just left me on his way to home. I am sending along $150,000 of gold with him. Boys, I wish I could say to you, there it is boys, take it and good luck to you. If I did that R. [Russell Burrows] would be up the creek. I have a lot for him to do yet, and the only way he has of getting by is the way we are doing it now. Everything is going to depend on how much he gets done in that hole, so boys, I hope you

Fig. 8.6. These gold medallions from the cave feature the same two Numidian letters for S and D, but in the coin on the right they are reversed. Photographs by Beverley Moseley

can hang in and bite the bullet. Look at the sunny side of the mountain, though, you are making one hell of a profit.[10]

To the readers of *Ancient American* magazine, in which this information was made public, the tenor of such correspondence implied fear of the U.S. Internal Revenue Service. The landlord's desire to remove "as much gold as possible" from the cave before the annual appearance of "some problems," and the mention of Burrows being "up the creek" unless some unspecified procedure was followed by all concerned, suggested the concealment of taxable gold.

Jack Ward was less inclined to conceal anything, and wrote to Frank McCloskey for assistance in legally disposing of their hoard. The Indiana congressman replied on March 31, 1989, "Thank you for contacting my office concerning your questions about gold. My office has been in contact with the U.S. Department of the Treasury concerning your questions."[11] McCloskey referred him to the U.S. Mint in Washington, D.C., which "buys gold directly."[12] Even Alexander P. MacGregor, a professor of classical history at the University of Illinois and severely critical of the cave's existence, remarked, "Now we know *this* gold is real, because Ward asked his congressman how he could go about selling gold to Fort Knox. The obliging congressman replied and his letter is reproduced in this issue (volume 3, number 16) of *Ancient American;* we have proof that Ward wrote the letter."[13]

Fig. 8.7. Most of the gold medallions from the cave are not currency but rather commemorative medallions, like those at left, minted to celebrate the official opening of the Gold Pyramid House in Wadsworth, Illinois, in 1977. The cave medallion at right, for instance, might memorialize the Mauretanians' transatlantic voyage from West Africa. The figure at the stern may be paying homage to the sun god, and appropriately the glyph of sun king Alexander Helios appears at the bottom. Photograph by Beverley Moseley

When these revelations were published in 1997, Russell Burrows did not sue Harry Hubbard for libel, as some expected and even hoped he would. Hubbard went on to accuse Burrows of "looting the cave," which he was using as his own private trove, hoarding its gold, and of selling off its thousands of artifacts for hundreds of dollars each. Burrows refused to respond, maintaining instead his original story, and continuing to insist that every ounce of gold taken from the cave had been returned. Although Hubbard believed otherwise, Burrows's honesty and integrity did not lack champions. His brother-in-law Sam Eyer, a respected schoolteacher, declared:

> I would like to say on behalf of Russ Burrows that I witnessed and initially helped with the uncovery of the cave and area around [it], which is in question. There at the cave site I personally saw what [were] undoubtedly ancient artifacts. I would further say that Russ has always been honest with me and I don't believe he would try to deceive anyone on that matter. I know for a fact that he has spent many hours of study on his findings and personally consider him ... learned in archaeology. I would not write this letter if I had any doubt in my mind about these findings.[14]

Five years later, John Tiffany, editor of *The Barnes Review,* reported:

> It is claimed that Burrows sold off enough artifacts to unknown buyers that he was able to place seven million dollars in Swiss numbered bank accounts. According to Swiss journalist Luc Buergin, this money derives from the illegal sale of gold artifacts obtained at the North American burial site. (Other sources claim that Burrows melted down all the gold and sold it as ingots. Still others question whether there ever was any gold in the first place.) Buergin accuses Burrows of having clandestinely sold thousands of "burial gifts." In his recently published book, *Geheimakte Archeologie,* he presents documents, financial papers, and pictures which indicate that Burrows has removed enormous quantities of gold from the cave system.[15]

In the book's appendix, Buergin reproduces a five-page handwritten itemization describing gold ingots, coins, breastplates, bracelets, ankle bands, neck ornaments, and headpieces, all allegedly removed from the cave, totaling in value $19,629.86; $3,029,789.62; $3,965,518.50; $5,821.81, and $13,580.00, respectively. Impressive as these figures are, they represent only a very partial cataloging of the complete hoard collected for the Artifact Recovery Exchange. In the opinion of investigators like Buergin, the very words chosen to define their enterprise defined the real intention of Burrows, Ward, and Cook—namely, to exchange the recovered artifacts for cash.

9
Find or Fraud of the Century?

> *They couldn't possibly come from Illinois!*
> OSCAR MUSCARELLA, ASSISTANT DIRECTOR,
> HISTORICAL DEPARTMENT,
> THE METROPOLITAN MUSEUM OF ART

While the gold was unquestionably real enough (*too* real to handle without perhaps running afoul of the Internal Revenue Service), most people with at least some knowledge of antiquities dismissed the Burrows Cave stones as crude fakes. Even cultural diffusionists, who advocated the existence of overseas visitors to the Americas before the arrival of Columbus in the New World, regarded the objects as elements of a transparent hoax. Such a conglomeration of various written languages mixed with what appeared to be Roman, Phoenician, Hebrew, even Christian "artifacts" and faces portrayed in stone was impossible to accept as genuine because it fit no conceivable ancient world or pre-columbian context.

Of course, Burrows and his colleagues lacked any professional credentials to establish the authenticity of items removed from the cave. So, at their request, Dr. Robert Pickering, an archaeologist from Chicago's

Field Museum of Natural History, visited the Sonotabac Museum, where he examined some of the controversial stones in April 1989. He said they merited further study, and returned with several specimens to Chicago. There he examined them together with his colleagues Bentram Woodland, curator of Petrology; Frank J. Yurca, a Ph.D. candidate in Egyptology at the University of Chicago; Catherine Sease, Field Museum conservator; and David S. Reese, research associate at the Oriental Institute at the University of Chicago. Dr. Pickering mailed the Burrows Cave specimens back to Ward the following month with a brief letter: "I see no indication that these artifacts are old or authentic. The most likely explanation is that they are modern fakes perpetrated by someone with a superficial and jumbled knowledge of the ancient Old World."[1]

Dr. Yurca was also unimpressed. "Stylistically, there is nothing in these objects to link them with authentic ancient Egyptian antiquities seen by me in Egypt and in eighteen years of the study of Egyptology." He referred to one of the portrait stones as "utterly un-Egyptian," with hieroglyphs that are "utter nonsense, not one single sign being a standard Egyptian hieroglyph or even an attempted copy of one." He went on to say of the figure depicted, "The arm and head are also virtual parodies of Egyptian art." He concluded thus:

> In summary, not one of these reliefs on marble looks the least bit Egyptian, let alone Old Kingdom. In my opinion, they are crude imitations. Iconography is garbled; symbolism is confused; supposed hieroglyphs are gibberish. The carver clearly had a vague, passing acquaintance with Egyptian art, but no understanding of it. The supposedly basalt pieces that I was shown were even more utterly un-Egyptian in style. Some of the faces had almost New World (Mexican) appearance, and some of the iconography (for instance, faces on the ends of scepters or weapons) are utterly un-Egyptian. In all the faces carved on these pieces, both the supposed basalt and the supposed limestone, I detect the workmanship of a single hand, suggesting that these are all the output of a single individual who shows a woefully inadequate or utter misunderstanding of Egyptian art and symbolism, as well as hieroglyphs.[2]

Field Museum curator Catherine Sease said of the Burrows Cave stones, "I must admit that I have never seen anything like them."[3] Dr. Reese's reaction was the same. "I have never seen anything like these objects from a published excavation or a museum. I am very certain that they have recently been made—their crudity, the stone used, and the figures and inscriptions present are possibly based on some real objects, but others are obvious fakes. I hope that the purchaser of those objects did not pay too much for these forgeries."[4]

Burrows was shocked and saddened by the Field Museum experts' condemnation of his artifacts. "I just couldn't understand it," he confessed.

> It didn't make sense. I knew that I was not manufacturing these things, and I knew that Ward was not manufacturing them and hiding them in the cave for me to find. I felt like someone had knocked my legs out from under me. Where were we now? It sure looked to me like we were out on a limb, and it was being sawed off. What should our next step be? It was obvious that we were beaten. We had spent a lot of time and money, and now we were stuck with a lot of worthless artifacts. Ward, Cullen, and I talked it over. How could anyone have faked those stones? We decided we should get back the tablets that we had sent to Chicago, and try to figure out how we had all been fooled.[5]

But Ward was less depressed than annoyed by the harsh judgments of the Field Museum experts. He sought out the state archaeologist, Mark E. Sary, for his opinion. As a downstate regional archaeologist more personally familiar with Illinois antiquities, Ward believed he could not fail to recognize the cave finds as legitimate and vitally important artifacts. Russell Burrows brought along a dozen of the illustrated and inscribed stones to the Springfield office. But Sary, speaking bluntly on behalf of his colleagues, dismissed the cave investigators and their alleged finds: "We have no interest, as long as they stay away from burials and burial artifacts. We believe it is a hoax or a practical joke."[6]

Although Ward felt denigrated by someone from whom he sincerely sought assistance, his meeting was not without value, because whatever doubts anyone might still have harbored about Burrows's honesty were dispelled by his visit to Sary. Burrows may not have been a Rhodes scholar, but he never would have been stupid enough to try to palm off fakes on the state archaeologist in person.

Yet, how could Sary, with his scientific background and years of experience, have brushed aside the objects as "a practical joke"? The man's expertise was, after all, confined to the tedious examination of native pottery shards and flint arrowheads only a few centuries old. Nothing in his experience really prepared him for ancient Old World artifacts in Illinois.

Apparently, Sary did not tell his superior anything about Burrows Cave. In a letter dated October 19, 1992, Thomas E. Emerson, chief archaeologist of the Illinois Historic Preservation Agency, informed Lois Benedict, director of the Institute for Archaeological Research, Inc., in Michigan:

> [A]s far as I have been able to determine, the individuals involved in this project have made no attempt to contact any members of the state's professional archaeological organization, nor has there been any contact with this office, with the exception of several telephone calls during the last year from Mr. Russell Burrows. Mr. Burrows briefly informed me of the possible existence of Burrows Cave, intimating that he had little involvement with it, and requesting information on the state burial law.
>
> The essence of our conversations was that the state law was designed to create an orderly process for the treatment of graves that were not in designated cemeteries. All the excavations must be carried out by a professional archaeologist, using approved professional standards, and unless claimed by appropriate related groups, the skeletal material and associated grave artifacts are deposited in the Illinois State Museum. No one has ever submitted a permit application to this office for Burrows Cave. It should also be noted that failure to report the disturbance of protected graves, as well as the actual disturbance of such graves, are both criminal offenses. Consequently, if

any activities were being carried out at Burrows Cave that disturbed human graves, they would be in violation of the Act. It is my present opinion that there is little evidence to support either the existence of any protected burials and/or associated artifacts at a location known as Burrows Cave. Should evidence be produced to indicate that the illegal disturbance of protected burials is occurring, our agency would notify the local law enforcement authorities, and pursue prosecution of those involved.[7]

Doubtless, the moment he got off the phone with Emerson, Burrows swore never to reveal the cave's whereabouts to anyone for fear of having all his finds, to say nothing of himself, locked up by the State of Illinois. As long as the site was known to him alone, he had little to fear. His sense of security was heightened by the steadfast opposition of professional scholars to his discovery.

The mind-set that dominated his thinking was dramatized in a meeting between Virginia Hourigan, a photographer of the Burrows Cave artifacts, and Oscar Muscarella, whom Hourigan described as "a genius in authentication at the Metropolitan Museum of Art."

According to Hourigan, after he glanced through her photo collection of the portrait stones, Muscarella said, "Well, I can see you're committed to this. And I'm really sorry to have to tell you that it's all a fake." Hourigan told him not to be sorry. The rest of their conversation went something like this:

"Please tell me what it is about them that gives them away."

"Nothing."

"Nothing? What do you mean? Surely, something leaps to your experienced eye that I wouldn't notice. What is it?"

"I can't tell without examining the stones in my laboratory, and I haven't time to do that."

"But if you don't see anything fraudulent, how can you tell they're fakes?"

"Because they couldn't possibly come from Illinois. If you said California, I might think it barely possible. But Illinois? No way!"[8]

Muscarella was in lockstep with most professional archaeologists who were firm in their conviction that no overseas visitors of any consequence landed in America before Columbus. Even after it was determined that a Viking settlement dated to 1000 A.D. had been established at Labrador's L'Anse aux Meadows, they refused to concede any further possibilities for Norse or other pre-columbian arrivals. Dr. Gunnar Thompson, a researcher at the ancient ceremonial center of Cahokia, in western Illinois, explained that "most historians and anthropologists are loyal to a doctrine of cultural isolation that was originally promulgated by a medieval religious fraternity. During the 1800s, the Columbian Order promoted the ethnocentric belief that Columbus was chosen by God to bring the first Christian civilization to America." Thompson continues:

> Although modern scholars abandoned the religious premise of American discovery, they adhered to the belief that no significant voyagers preceded Columbus to the New World. This belief is often referred to as the Monroe Doctrine of Cultural Isolation. Because of this doctrine, establishment scholars automatically dismiss evidence of pre-columbian cultural diffusion as heresy. The resulting academic myopia is a clear indictment of scholars who claim that their beliefs are based on scientific principles. Indeed, the practice of science demands an open mind and examination of all the evidence no matter how unorthodox it might seem.[9]

Thompson concludes that "America's institutions of higher learning have promoted blind obedience to a medieval religious doctrine. So pervasive is this academic prejudice that professional careers have been ruined when individual investigators questioned prevailing scientific dogma."

The renowned Dr. Michael Coe, among the most important mesoamericanists of the twentieth century, found and first studied the Olmec Civilization, America's earliest. He is universally considered the leading authority on this seminal culture that flourished in the Mexican states of Veracruz and Tabasco from 1500 B.C. to 800 B.C. Yet, Dr. Coe admitted from the safety of retirement that if he had publicly discussed his observations of Indonesian influences in mesoamerica, his days as a practicing archaeologist would have been terminated.[10]

Blissfully unaware of the academic orthodoxy dominating American archaeology, Jack Ward continued to make the rounds of university-trained scholars, who mostly dismissed his inscribed stones with patronizing smiles. Sometimes, however, they were able to move beyond denial. One of their nonconsenting colleagues was James P. Scherz (fig. 9.1), professor emeritus at the University of Wisconsin Department of Civil Engineering in Madison.

Fig. 9.1. Professor James Scherz, an early champion of the cave's archaeological authenticity

Years after he concluded that the artifacts were authentically ancient, he was asked to lecture with Burrows about the cave. "We were invited to give a presentation at the Falls of the Ohio Museum in Jeffersonville, Indiana," Scherz recalled.

We had given similar talks before, and I merely gathered my slides and overheads [projection transparencies] to make ready for the drive down. About forty people had assembled in one of the buildings at the park. Our host was a young director of the facility who worked for the state parks and the Department of Natural Resources. As the time for our presentation was nearing, we were chatting together, when he got a phone call which obviously both annoyed and amused him. After hanging up, he said sheepishly that he had just heard from the state archaeologist, who told him that no lecture about Burrows Cave would be allowed in any state building. "Why?" our host [had] asked.

"Because all of us in authority for such things know that the Burrows Cave stones are modern fakes." "Have you seen any of them yourself?" he [had] asked. "No. But we know all about them. Nothing like this has ever been found in the New World, and they have to be fakes," was the authoritative reply. Our host said that did not sound very scientific; he still wanted to hear our lectures, and to see some of the rocks himself before he made up his own mind.

He rented a meeting room in a local motel, where we gave our presentations, and members of the audience were able to handle the stones Burrows had brought along. But all got a good education about the standard method used by such state-appointed officials, to whom we entrust the ancient history and prehistory of our land, as they decide which data to label "authentic" and which are modern forgeries. If new data is reported which fits what they already believe, then it might be authentic and worth investigating. If reports of new data appear to be different than what they have already seen and believe in, then the unfamiliar materials are obviously fake, and there is no need for consideration and examination, whatsoever. The standard policy used by many of our appointed authorities is based on dogmatic beliefs, rather than on a fair analysis of basic data, as required by the scientific method. It is extremely difficult for a person well indoctrinated into a dogmatic belief to begin questioning that belief, even by the scientific method. It is as difficult as tearing down a large wall made of bricks and concrete which their honored predecessors had laboriously built. It is no simple or easy undertaking.[11]

In other words, the official archaeologist of Indiana, a state from which the inscribed stones were not taken, forced from a public facility the impartial host, compelling him to rent a room at his own expense, because the archaeologist condemned the artifacts without ever having seen them. To Scherz, such arrogant prejudice was not only antithetical to the spirit of freedom of expression, but certainly violated the fundamental principles of scientific inquiry itself. Scherz was, however, encouraged by a few fellow professionals who likewise dared to disagree with the majority of their colleagues. Dr. Curtiss Hoffman, professor of anthropology at Bridgewater State College in Massachusetts, was skeptical but at least willing to consider the possibility of the stones' ancient authenticity. He recommended a series of alternative hypotheses any research program should consider in order "to achieve the only one standing." These hypotheses included:

1. The site is an undisturbed burial of Old World cultural

affiliation, dating prior to the discovery of the New World by Columbus.
2. The site is an undisturbed burial of Old World affiliation, but dates later than the arrival of Europeans into the area.
3. The site is a burial of New World cultural affiliation into which Old World materials have been introduced.
4. The site is a hoax perpetrated by some person or persons knowledgeable of Old World scripts and practices, who produced the material and introduced it into the site.
5. The site is a hoax perpetrated by a person or persons having access to genuine Old World cultural materials, who introduced these to the site.
6. The site is an undisturbed burial of New World cultural affiliation whose contents have been mistaken for Old World materials, or are natural, unmodified items.[12]

But the site upon which verification of the hypotheses depended was known only to Russell Burrows, who steadfastly refused to disclose its location. In 1986, James Whittal, of the Early Sites Research Society in Massachusetts, whose members were professional as well as avocational archaeologists interested in American prehistory, composed a methodology for following up on Hoffman's suggested alternatives. Their ten procedural points involved the creation of a database for all material relative to the site, together with samples of metal and lithic artifacts to undergo nondestructive analysis. "The cave must be seen by an outsider who has archaeological expertise, and probably a good geologist, as well," he concluded. "At the time of observation, the event should be videotaped."[13]

Burrows responded to these eminently rational suggestions in a letter to the ESRS membership by declaring, "Not one of you will ever see this cave, and that is a statement that I will abide by for as long as the grass grows, the wind blows, and the rivers run." He dismissed Whittal's proposed methodology as wholly unacceptable, and refused assistance from the society, which he never solicited for input of any kind. But he did publicly promise to escort one highly respected scholar into the cave himself.

Dr. Cyrus Gordon (fig. 9.2), professor of Mediterranean studies at Brandeis University, was a world-class Semiticist until his death in 2001. His expertise was matched only by his courage in defying academic orthodoxy, as he illustrated when he publicly identified a stone text found in Brazil (the Parahyba inscription) as an authentic artifact describing a crew of Canaanites from the Phoenician city of Sidon arriving on the shores of South America in the sixth century B.C. He gained further renown for his verification of the Bat Creek Stone, found by Smithsonian Institution archaeologists near Knoxville, Tennessee, in 1885.

Typically dismissed as fraudulent by establishment scholars, the inscription of this stone is a Hebrew text dated by Gordon to the second century A.D. His conclusion, based entirely on the Bat Creek Stone's internal linguistic evidence, was independently confirmed by radiocarbon analysis of some wood fragments originally found with the tablet. In 1989, Smithsonian laboratory technicians, using carbon-14 testing, determined that the artifact had been buried between 32 and 769 A.D. Clearly, Gordon was open to the possibly ancient authenticity of the Burrows Cave stones. His corroboration of their overseas provenance would have enormous impact on the whole controversy. In June 1991, he met Russell and was shown the photographic slide of a gold coin allegedly taken from the cave.

Fig. 9.2. Dr. Cyrus Gordon tells a meeting of the Institute for the Study of American Cultures about Burrows Cave.

"I felt that the coins, if genuine, were important enough to merit investigation, rather than be brushed aside, *a priori,* as 'too good to be true,'" he wrote the following September.

> I remember when the caves that yielded the Dead Sea Scrolls were considered "too good to be true." However, until a trustworthy and competent scholar is permitted to examine the actual coins, we must countenance other scenarios, such as that the slide could have been

made from a recently contrived painting. No credible witness has, to my knowledge, seen the cave or the coins.

The descriptions of the cave and its contents that have come to my attention are topologically akin to tales of caverns in the *Arabian Nights*. I have also heard such "cave stories" from sincere natives during my years of excavating in Iraq and Jordan. But Illinois is a new and unexpected source thereof. I am interested in the phenomenon of belief in the fanciful. Perhaps all of us are subject to this to some extent. But it has no place in the disciplines of archaeology and history. If Mr. Burrows is essentially sincere about the cave as he has described it, he may be obsessed with fantasies like many of my Near Eastern acquaintances who live in a wondrous world full of *jinn* and *afarit*. Their supernatural experiences are full of superlatives, like Mr. Burrows' tales.

The supposed owner of Burrows Cave is said to have allotted twenty-five million dollars for excavating the cave and building a museum to house the contents. The cave is said to include thirteen tombs with human remains, a ton (sic!) of gold, and even inscribed scrolls. Mr. Burrows tells how he penetrated the cave by stepping on the outside half of a loose slab at its entrance. Had he instead stepped on the inside half, he would have dropped to his death in a pit made to trap intruders, as in pharaonic tombs in the Valley of Kings. If only a tenth of Mr. Burrows's account were true, the cave would be an outstanding discovery. So, at the ISAC [Institute for the Study of American Cultures] meeting in [Columbus, Georgia], June, 1991, I arranged to examine the cave and the gold coins under the guidance of Mr. Burrows. Soon, we fixed an exact date: October 19 and 20, 1991. After a month and a half had elapsed, I received a couple of phone calls from Mr. Burrows. In the second one, he asked if I was up to walking about half a mile to reach the cave from the road where we would have had to leave the car and continue on foot. I assured him that I would be equal to it (for I walk more than that every day).

Then he added that the area was infested with snakes, to which I replied that with high, leather footgear, I had no objection to the presence of serpents. (I'm used to venomous creatures from my excavations and explorations in the Near East, where I regularly shook out my shoes to make sure there were no scorpions in them, or when I got dressed in the morning.) When his latter-day Labors of Hercules failed to dissuade me, Mr. Burrows notified me (in a letter dated 23 August 1991) that the owner had rejected our long-planned visit to the cave site, for security reasons. However, Mr. Burrows courteously offered to show me some of the region in which the cave was allegedly situated, and his collection of 115 artifacts from the cave. I declined this kind offer, because I have examined some of those finds and seen the photographs of many more. They are modern products in which I am not interested. The "good stuff" (notably, the gold) have—so the story goes—been put back into the cave.

I cannot prove that the cave, its owner, the latter's munificent gift of twenty-five million dollars, and the ton of gold do not exist. In a world where men walking on the moon have been televised to Earth; where surgeons perform heart transplants; where the Berlin Wall came tumbling down; where the Cold War suddenly ended, etcetera, who can say that other miracles can't happen? Maybe there is a tiny "kernel" of truth to the wild stories about Burrows Cave. But I have no time to search for it. I have other priorities.[14]

Gordon washed his hands of the whole controversy, even though, as he suspected, there was at least some truth to the cave story. How much, he could never determine, having been refused entry to the site.

Russell Burrows's version of the spoiled relationship is quite different. In his account, Gordon actually showed up in southern Illinois, and Burrows would have taken him to the cave, but the professor violated their previous agreement by bringing along his wife, Connie, who insisted on accompanying her husband. This the mysterious landowner refused to allow, and so called off the visit. Whatever really happened, Gordon's failed attempt to personally examine the site seemed to remove all hope for anyone other than Russell Burrows ever being allowed to enter it.

Its credibility was dealt another severe blow by Barry Fell. A marine biologist and cryptographer, Fell applied his professional background as a code breaker to examining alleged Old World inscriptions on pre-columbian artifacts. His books, *America B.C., Ancient Settlers in a New World,* and *Bronze Age America* popularized the cultural diffusionist cause by discussing evidence for alien arrivals in the New World many centuries, even millennia ago. In 1986, he condemned the entire southern Illinois collection as fraudulent after seeing the photograph of a single Burrows Cave stone incised with an inscription and the representation of an elephant at the top. It was almost identical to a specimen from Ecuador that Fell had been researching for at least the previous ten years (fig. 9.3). He had never actually handled it, but worked instead from the photo of a ceramic model of the South American artifact. All its letters save one— an "upright backward Z"— appeared to belong to an Iberic written language.

For his translation to work, it should have been a "lazy," i.e., backward, Z. He assumed the ancient inscriber had made a mistake, so Fell substituted his lazy Z for the "incorrect" Z and it was his substitution that appeared in the first release of *America, B.C.* He changed his mind shortly thereafter, however, and published in *The New England Antiquities Research Association Journal* (no. 19, 1990) a drawing of the Ecuadoran elephant stone with its original upright backward Z, because its lazy alternative could not complete his own rendering of the text. In the end, he was unable to translate it, and guessed, unconvincingly, that it described some gibberish about an earthquake-causing elephant! He was nonetheless certain that the stele, or stone slab, itself was authentic, however inscrutable.

Fig. 9.3. The so-called elephant stone from Ecuador's Crespi collection. The first letter of the top row is the "upright backward Z."

After his drawing appeared in the *NEARA Journal,* Fell was shown a photograph of the similar Burrows Cave elephant stone with the "lazy" backward Z he originally postulated, then discarded (see fig. 9.4). Photographer Virginia Hourigan, who sent him the print, reports that Fell, in a telephone conversation with her, was "irate and enraged. He would see to it that these fraud artists were caught and punished. He sounded as if he thought I was one. Finally, he announced that I was a very rude person, and slammed down the phone."[15]

Nothing ever came of his efforts to "see to it that these fraud artists were caught and punished," although his published disdain for the southern Illinois artifacts did frighten away many people formerly interested in studying them. Fell's reputation as a difficult man to work with survived his death in 1991. Mortified by his failed efforts to make sense of the Ecuadoran elephant stone, he vented his frustration on the untimely appearance of the Burrows Cave look-alike.

Of course, Burrows himself was suspected of altering the glyph to match Fell's published interpretation. But according to Fred Rydholm, a retired Michigan high school teacher and one of America's most knowledgeable authorities on the ancient copper mines of the Upper Peninsula in the Great Lakes Region, Burrows "hadn't even heard of Barry Fell's books until he met Jack Ward in 1984, and by that time had already found the elephant stone in the cave."[16] While it seems impossible to solve the discrepancy between the glyphs of the upright *Z* and the lazy *Z*, it is very doubtful anyone altered the Burrows Cave version, something only Fell would have noticed, anyway. As it was, his "translation" of the script was useless.

But to at least some investigators, the entire vast collection could not be entirely dismissed as fraudulent. They knew that Russell Burrows and Jack Ward were not "persons knowledgeable of Old World scripts and practices, who produced the material and introduced it into the site."[17] The focus of their inquiry shifted from epigraphy to geology. According to John Whittal, who proposed the unacceptable methodology for researching the cave, "Jack Ward had a fifty-year career in the field of highway

Fig. 9.4. The Burrows Cave elephant stone with what appears to be the same inscription as that found on the stone in the Crespi collection

construction, traffic, aggregate production, and marketing. His specialty was prospecting for mineral resources. By his own account, he became thoroughly involved with geology, particularly with relationship to stones and their different characteristics."[18]

Most of the Burrows Cave specimens were black, discoidal stones, with which Ward was utterly unfamiliar. In his half century of quarrying and highway construction, he had never seen anything like them. But he did recognize the far fewer and completely different white stones covered with various written languages and the carved profiles of men and women wearing Egyptian-like headgear. "The slabs of white marble are numulitic limestone," he stated, "and appear to be the same type of limestone that was produced in ancient times from a quarry in the Mocottam Hills, on the east side of the Nile River, north of Memphis. Said location is southeast of present-day Cairo. These artifacts are of genuine prehistoric origin." With less certainty, he guessed that the black stones were "diorite and came from a quarry source in Nubia, which is in the Sudan."[19]

Dr. Norman Nielsen, a nationally recognized expert in the study of ancient metals and lithic material at DuPont Laboratories, concurred that white marble retrieved from the site was "of the same type" quarried by ancient Egyptians. But he did not agree with Ward that the tablets were necessarily "of genuine prehistoric origin," because numulitic limestone is found in various parts of North America as well, not only at the Nile Delta.[20] Ward's identification of the more numerous black tablets seemed incorrect, however. Diorite is a dark gray or greenish igneous rock composed mostly of feldspar and hornblende, a description that does not really match the Burrows Cave examples.

In December 1986, Dr. Nielsen received one of the cave's black oval disks for analysis. Cleaning a small area of the stone with ethanol, he and his colleagues at DuPont were surprised to observe that its surface grew suddenly dull, and a black stain came off on the polishing cloth swab. The stone's originally bright sheen reappeared with continued rubbing, suggesting to the scientists that it was "an artificial composite." Ever cooperative, Ward provided them with fragments from other black Burrows Cave stones, which were submitted to scanning electron microscopy, elemental analysis by emission spectroscopy, thermogravimetric analysis (burning), and infrared spectroscopy of a chloroform extract.

In a complete report of their investigation submitted to James Whittal, Dr. Nielsen wrote that laboratory testing revealed the black stones were mainly silicate with considerable calcium and iron. Heating during thermogravimetric analysis caused the loss of 14 percent of volatile material: water, two organic substances, and, finally, carbon. At 7,700 degrees Centigrade, the stone turned white, "due apparently to oxidation of carbon or other carbonaceous material. A chloroform extract was shown by its infrared spectrum to be an aliphatic ester." This is an organic compound belonging to saturated alloys formed by the reaction of acid and alcohol with the elimination of water. "Electron microscopy revealed that the fractured surfaces of the stone had a waxy, micronodular porous surface."[21]

Dr. Nielsen wondered if these results indicated that the Burrows Cave tablet was not natural, but had been artificially fabricated from stone dust, black carbon, and an organic binder of some kind. In August 1989, his data were sent to Dr. James L. Guthrie, an organic chemist widely experienced in polymer technology, to discover the identity of the binder. If so, new testing might determine if it was a modern synthetic material or something that would have been used by a human manufacturer in the ancient past. The stone's prehistoric authenticity or twentieth-century contrivance seemed to hang on the outcome: Was the tablet a natural rock or actually a man-made composite, ancient or modern?

Dr. Guthrie confirmed Nielsen's test results, although a geologist colleague believed the Burrows Cave stone was more a fragile argillite shale that had fractured along its natural cleavage planes. But argillite, a hardened mud stone, shows no slatelike cleavages. According to Whittal, "argillites are mainly hydrated silicates that may be black because of iron or carbonaceous inclusions, and they lose their color when the inclusions are burned out."[22] Dr. Guthrie observed small amounts of water in the porous rock, and found that its volatile organic substance comprised several waxes, mostly lanolin. Jack Ward told the scientists that he had used shoe polish and car wax on the stones to highlight their incised images for better photography.

In his efforts to dissolve the polymeric binder, thereby identifying it, Dr. Guthrie attempted extraction with eight different solvents, each stronger than the one preceding it. But even the application of the most powerful dissolver (dimethalformamide) brought about no physical appearance or weight loss, and failed to bring out any substances, other than 4.5 percent and 0.3 percent residues of Ward's waxes and polishes. Laboratory testing of the Burrows Cave lithic materials showed they had no binder, ruling out any possibility that they were man-made in either ancient or modern times. The black tablets were definitely argillite, a soft shale commonly found throughout southern Illinois. While the true age of their inscriptions and portraits could not be determined, nothing about the stones indicated they were made in modern times.

Meanwhile, following up on a related lead, Jack Ward related that he:

> went to Livingston, Kentucky, and talked to a Mr. Carlos Coloftis, who I learned had similar artifacts [to those of Burrows Cave] on display at his Indian trading post, which is known as Fort Sequawyah. Mr. Coloftis showed me a similar array of artifacts that he claimed were found in a cave in Kentucky along the Virginia state line. He informed me that the Kentucky cave had a lot of the black-type artifacts. Prior to the development of the Carlyle, Illinois reservoir, about thirty miles west of Burrows Cave, an archaeological salvage dig was performed for the U.S. Army Corps of Engineers. Excavators selected two sites to be investigated, one near the south side of the proposed lake, the other on the north side. The archaeologists found nothing in their south pit, but in the north pit they found over *350,000* (!) flint blades which they termed the "North Blade Type." The shape of this type is very similar to the black picture stones [from Burrows Cave]. The archaeologists determined that the North Blades were funerary items for trade.[23]

Ward's personal investigations showed that Burrows Cave–like artifacts, as well as the mud stones from which they were made, occurred beyond the Olney site, even outside Illinois, revelations that underscored its archaeological authenticity. Later, however, Wisconsin antiquarian Wayne May, while researching the suspected vicinity of the cave, was surprised to find hundreds of mud stones identical to the kind retrieved by Russell Burrows. They were "scattered everywhere and exposed in strata along the river bank," he said. Seeing them in such profusion, he rejected Ward's opinion that they were originally brought over from the Old World as ballast in the hulls of ships sailed by the men who later incised the black stones with scripts and images. Referring to the engraved tablets, May observed, "[T]he material for their manufacture was close at hand, as it still is today."

But reproducing a Burrows Cave–like artifact from one of the virgin argillites was very rough going. "Marking with a nail and screwdriver was difficult, while the stone chipped and flaked at every stroke," May found. "Clearly, the rock face was impossible to work" if an artist or engraver wanted to create the same smooth lines of the tablets' gracefully rendered portraits of men and women or the precise and intricate characters of a written language.[24] The brittle exterior simply did not allow for such artistry, regardless of the tool or implement used to execute the images and glyphs.

How, then, were they formed with an apparent ease that facilitated their creation by the thousands? The answer to this question, at least, was apparent. Black clay with high iron content, such as is found along southern Illinois rivers, may be carved out from riverbanks in oval chunks with a paring knife or similar tool. These chunks, while still moist, are very malleable and can be handily molded into any form. But after being smoothed on one side, they particularly lend themselves to writing inscriptions or making illustrations with a stylus, and to the creation of impressions in the soft clay with a stamp, signet, or font. If the chunk is left exposed to dry air conditions, it eventually turns into hard mud stone, or argillite.

How long such a process takes is unknown, if only because different-sized chunks with varying amounts of iron will not dry out at equivalent rates. At least several years must pass, however, before the clay is hardened all the way through for an average-sized Burrows Cave stone—probably too long for a modern forger to wait before he could sell it. Tests of all black stones from Burrows Cave showed no signs of heating or baking to induce petrification. But if argillite could be made to lose its hardness, then it might become soft enough again for a present-day forger to inscribe the surface with glyphs or illustrations.

As an experiment, May soaked an unworked specimen in water. After four months, the mud stone did not return to its former malleability. On the contrary, it began to dissolve and disintegrate. Clearly, it could have been worked with a stylus only while still clay. After it hardened into argillite, restoring its claylike plasticity was not possible.

In January 1997, I took one of the smaller Burrows Cave artifacts to gemologist William Wild (fig. 9.5), an experienced lapidarist and professional artist well versed in Native American cultures, for testing at his jewelry store in Orland Park, Illinois.

The item chosen for his examination was the oval black stone representation of a human face, 3.1 inches from crown to chin, 2.5 inches from ear to ear, and .51 inch at its thickest point. It weighed only 3.7 ounces. Wild submitted the Burrows Cave image to several nondestructive tests and observations, mostly using a 30-power illuminated magnifier. His observations confirmed at once that the carved face was unicursal (formed of one continuous line) and that its features had been subtly molded by the artist's fingers pressing into the soft clay.

Fig. 9.5. Lapidarist William Wild, who worked to establish the ancient authenticity of the inscribed cave stones

The image could not be reproduced on the same stone, which had become too hard for any similarly delicate rendering. It must have been made, Wild concluded, when the stone material was soft enough to allow for the face to be literally drawn with a stylus. Subsequent magnified examination of inscriptions on other Burrows Cave stones revealed none had been etched into the brittle argillite, but rather had been inscribed with a sharply pointed writing instrument on the mud stones while they were still moist clay.

Wild's attention was next drawn to the stone head's holes, one in the center of each ear. He noticed that they were made by drilling first one side, then the other, back and forth, until completed. He said that such a procedure had been typically employed by pre-columbian craftsmen in both mesoamerica and above the Rio Grande. Manufacturers of modern replicas would ordinarily drill straight through from one side to the other. He observed, too, that telltale marks on the inside of the holes were not made by a metal instrument, but appeared to have been scratched by a harder stone, implying its premodern manufacture.

The face likewise showed the same kind of marks resulting from work with a harder stone tool. Modern forgers of fake artifacts generally use steel implements. The ears themselves were unevenly wrought, an otherwise minor but in this case revealing point of comparison with Native American artistic conventions, which also depicted ears unevenly; the tendency of modern replica makers is to portray all facial features evenly. Closely examining the figure's mouth under the magnifier, Wild noticed that the outer ends of the lower lip were slightly turned up, a stylistic detail especially favored by Olmec artists.

The Olmecs created America's earliest known civilization, beginning around 1500 B.C., at Vera Cruz, eventually spreading their high culture across Mexico to the Pacific shores of Oaxaca. Wild admired the general molding of the face, observing that its sculptor had achieved some realistic effects through subtle polishing of the stone to suggest more than actually depict a convincing chin and cheekbones.

Perhaps the single most enlightening piece of evidence found in Wild's examination belonged to the patina found on the sculpted elements themselves. Of course, the carving could have been made only before incrustation or calcification took place, and the rate at which patina is laid down depends on many factors, but generally the process involves many years, depending on its accumulated thickness. Patina can also be faked, although to achieve a convincing appearance takes skill. The carved stone could be a modern fabrication not only if the forger were gifted in original mesoamerican artistic styles, down to some very small and otherwise insignificant details, but also if he were in possession of and skilled in using the same kind of tools used by the ancients, and if he were adept in coating his fraudulent piece with a most persuasive-looking patina.

To be sure, all this could be achieved today by a knowledgeable and skilled forger after going to substantial trouble, but only if the work was intended to pass close scrutiny by experts to determine its genuinenesss. In such situations, important financial questions are always at issue. But this obscure Burrows Cave item, at least, already had been privately owned for three years, during which time (as well as before) no attempt was made by anyone to promote acceptance of its authenticity. Wayne May received it as a gift from Russell Burrows. No money was ever exchanged for the stone face.

Mr. Jesse Myers, who owned a stretch of land along the Big Creek River near Calhoun, Illinois, not far from the suspected location of the cave near Olney, told *Ancient American* magazine that he often gave Russell Burrows permission to pick up specimens of argillite, which may be found in profusion there. Those who deny the cave's archaeological authenticity speculate that he personally etched some of the cruder line drawings and script in these available mud stones when his stash of real artifacts began to run low, then palmed off these fabrications as authentic pieces to gullible buyers.

While these skeptics could never prove their assertion, why Burrows would have bothered collecting unworked argillite is an unanswered question many investigators of his discovery still find disturbing. In his defense, he claims he knew nothing about argillite, but was instead fossil hunting along the Big Creek—and, in fact, argillite frequently contains fossilized materials.

Heartened by the results of such testing, especially by Drs. Nielsen and Guthrie, a few professionals renewed their interest in the site. Then Beverley Moseley, president of Ohio's Midwest Epigraphic Society, in Columbus, noticed a small crystal growing inside the groove of an Ogham inscription on one of the Burrows Cave stones he purchased from Russell Burrows (see fig. 9.6).

Moseley showed it to a local geologist, who told him that the crystal needed more than one thousand years to have attained its present size. The inscription had to have been made before then. To Moseley

Fig. 9.6. Growth of a crystal embedded in one of the southern Illinois stones (near right edge of photo) confirmed it had been carved more than a thousand years ago. Photograph by Beverley Moseley

and other investigators, here at last was firm geologic proof that at least one of the controversial tablets was authentically ancient.

Also impressed was Dr. Warren Cook (fig. 9.7), professor of history and anthropology at Castleton State College in Vermont. Traveling to Maine, he conferred with Dr. George Carter, a geographer at Texas A & M University, for the purpose of organizing an archaeological team to investigate the Burrows Cave phenomenon from every scientific angle. A multifaceted scholar fluent in German, French, Spanish, and Portuguese, Cook earned his first doctor of letters at the oldest school in the Western Hemisphere, Peru's Universidad Nacional Mayor de San Marcos, in Lima.

From there, he went on to obtain his M.A. and Ph.D. at Yale University, where his doctoral dissertation, entitled *Floodtide or Empire: Spain and the Pacific Northwest, 1544–1819*, won the coveted Herbert Eugene Bolton Prize for the best book of

Fig. 9.7. Dr. Warren Cook was the first investigator of the cave artifacts to suggest their North African origin.

1973 in Latin American history. Since 1958, Cook had been renowned as an authority on the Spanish of the Far West. Although crippled with polio from childhood, his on-site research of pre-Conquest cultures in the Peruvian Andes won him the international acclaim of his academic peers, who marveled at his expertise in determining the authenticity of ancient artifacts.

Cook wrote of himself:

> For twenty-eight years at CSC [Castleton State College] I have seen my primary function as a teacher to demonstrate to my students, by example, how important it is to challenge the "experts" when their pontifications don't account for all the known facts. My proudest accomplishment in life thus far is the feedback testifying to the fact that I have been able to transmit this skeptical approach to thousands of Castleton students, who have gone on to plant this same attitude in other places, in many disciplines. I see myself as a Colombo [sic] in rumpled raincoat, the ethno-historian as detective. What really happened? Why did people do what they seem to have done? What are the real reasons for cultural change?[25]

Of all the university-trained professionals to investigate the southern Illinois controversy, Cook was perhaps the best qualified. Given his eminent background, scientific credentials, and particularly his rare open-mindedness, his elementary interest in the cave went far to champion its prehistoric credibility. Indeed, he was an outspoken proponent of its retrieved objects as legitimate artifacts. His keen powers of observation allowed him to distinguish easily enough between pristine evidence and evidence that had been tampered with.

Russell Burrows and Jack Ward were naturally honored by Cook's enthusiastic support of their efforts, and they invited him to join their Artifact Recovery Exchange as its project director. He threw all his energies into not only establishing the site's authenticity beyond question, but also unraveling the mystery of its ancient inhabitants, including where they came from and when. All the tablets in the A.R.E.'s collection were placed at his disposal, although even the prestigious

Cook was not allowed entrance to the cave. He good-naturedly set to work, however, studying the numerous inscriptions with their puzzling profiles of strangely attired men and women. After only a few months, he was ready to conclude:

> Burrows Cave's carved and inscribed stones are not modern forgeries. There is no doubt but what the many hundreds of stones would necessitate thousands of hours of skilled incising by a veritable platoon of inscribers, all at home with a style of line, letter, and a cultural, mental and physical imagery foreign to modern forgers. With several notable exceptions, most are singular artistic compositions, not directly copiable from known illustrations of ancient relics. They are often executed with great beauty, and, in my opinion, skilled epigraphers—of which there are very few in America—will eventually extract meaningful translations from many of them.
>
> Forging them would be virtually impossible, even for a specialist. There is evidence that the inscriptions are found elsewhere in the region, besides Burrows Cave. Of special importance are the pipe stems found recently (ca. 1987) near a Vincennes, Indiana, golf course, bearing comparable inscriptions, demonstrating that the people of the burial cave influenced the local inhabitants of the area. Never before has so much artifactual material turned up in one site that so radically challenges the present paradigm—that is, mind-set—of ancient history. The reigning dogma in textbooks is that there was no substantive transoceanic contact or transfer of cultural or genetic material prior to the Vikings and Columbus. Virtually all scholars whose education has solely been in American universities were taught this was the case. Their almost unanimous response to material such as Burrows Cave is: "It cannot be true, therefore it must be false, and therefore be either a misinterpretation or counterfeit."
>
> I consulted [July 3] with Dr. George Carter, distinguished professor emeritus of geography and archaeology, Texas A & M, author of *Earlier Than You Think,* dean of specialists in the study of diffusion, and one of the founders of the Epigraphic Society. What Carter finds most persuasive is the very volume of artifacts thus far extracted,

which, from my description, would be impossible to counterfeit. He made the following recommendations: A committee of specialists should be contracted to do the dig. Finding the chief archaeologist he sees as posing the biggest problem, because most of them are so close-minded and biased against recognition of any evidence favoring transoceanic diffusion. Nevertheless, he recommends one man, a distinguished archaeologist from the University of Illinois. Carter himself would be interested in being a member of said committee, and he recommended inclusion of a distinguished lady Ph.D. in linguistics (an acquaintance of mine) to take up the challenge of identifying and translating the various ancient scripts involved.

In my opinion, Burrows Cave ranks as one of the most important discoveries ever made in North American archaeology. I feel that the Burrows Cave site and material is so abundant and epigraphically challenging that it merits being a focus of laborious study for me and others surely the rest of my scholarly life.[26]

10
Fire in the Hole

We should welcome skeptics willing to test instead of burn Copernicans at the stake.

DR. CYCLONE COVEY

Unfortunately, before Dr. Cook could develop and present his research supporting the cave site's overseas origin, he passed away in his sixty-fourth year, less than three months after coming to his earthshaking conclusion. His death was followed just two years later by the passing of Jack Ward at seventy-nine years of age, in 1991. The loss of these two men was compounded shortly thereafter when Norman Cullen, one of the three original founders of the Artifact Recovery Exchange, died suddenly.

But even prior to their untimely passing there was dissension in the ranks. During late 1988, Russell Burrows gave 1,993 artifacts to Jack Ward for storage at Indiana's Sonotabak Museum. "The agreement was that he and another gentleman [Cullen] in Vincennes would cover the cost of the cave," Burrows explained, "and for that I would place all recovered artifacts in the museum." But the following February, his inventory of the cave materials there amounted to only 445 objects.

Two months later, he could find just 356 pieces. "On the night Ward died," Burrows said, "I went to Vincennes and recovered 117 artifacts with Mrs. [Mildred] Ward's consent and in the presence of three witnesses. That leaves 239, which, as I understand it, are in the possession of her attorney [Robert McCormack]."

Burrows recalled clearly an agreement—in writing, he asserted—that he, Ward, and Cullen made regarding the selling of artifacts in the interest of raising funds: Each of them was to receive a third of the proceeds. Burrows insisted, "I was never given a cent from artifact sales. As a matter of fact, Ward insisted that he had never sold them. Ward also liked to tell people that I owed him a large sum of money from loans and advances. I have receipts which prove that I repaid every cent he loaned to me, as well as advanced, for work in the cave."

Burrows also asserted that, at his insistence, the Artifact Recovery Exchange was dissolved in 1989, after an IRS audit of Ward's finances resulted in a claim filed against Ward. According to Burrows, who claimed to have in his possession the IRS agent's report, "He [Ward] told the agent that he had hired me to dig out *his* cave, and that what he had paid me was wages. Thank goodness that was cleared up by the appeals officer! After that, I told Ward and the other gentleman that there was no way that I was going to continue with them when they tried to hand me over to the feds to protect themselves."[1]

Burrows never explained what motivated him to make any kind of agreement with Ward and Cullen in the first place. Harry Hubbard charged that the sole purpose of Burrows's association with the A.R.E. was to fence the gold, diamonds, and artifacts he "looted" from the cave. In any case, of the original four researchers, only Russell Burrows was now alive. He continued to sell stones from the still-undisclosed site to a small circle of largely amateur antiquarians. A group of avid collectors in the Midwest alone is alleged to have purchased, over the years, an estimated one hundred thousand dollars worth of Burrows Cave artifacts.

Sales such as these brought down fresh contempt on the controversy, with charges from readers of *Ancient American* that Burrows, having looted the cave in the beginning, continued to make handsome profits by selling its irreplaceable artifacts, America's priceless heritage, to private buyers. Clearly, his critics said, his interest in the site and its archaeological treasures had been entirely venal from the outset. But he was not without his supporters.

"I know Burrows better than most people," declared Dr. John White, a retired physicist and head of Ohio's Midwest Epigraphic Society, in Columbus. " . . . We have many experiences in common. Some of the stories about Russell have a grain or two of truth to them, but they are frequently told by former Burrows Cave suitors or prehistory wannabes who have a personal ax to grind." Dr. White summed up his assessment of Burrows thus: "In a tough combat situation, he is the kind of man you would want to share a foxhole with . . . and based on my experience, I don't think he or his friends fabricated the Burrows Cave artifacts. He could have bought a truckload of phony artifacts from a stranger, but I don't know of any archaeologists with the money, time, interest, education, skill or inclination to get involved in such a huge enterprise."[2]

By 1989, the fabulous site had become the subject of much gossip from Olney-area residents. Some of them were making serious efforts to locate the camouflaged cave and loot its supposed treasures. In 1989 the Illinois state legislature was reviewing passage of a new law aimed specifically at protecting archaeological sites found by nonprofessional investigators. It imposed severe penalties, including jail time, on anyone who failed to report the discovery of prehistoric valuables to the authorities and sold the artifacts for personal profit. The legislators little knew or cared about the Richland County site but instead were concerned that the Slack Farm scandal might be repeated in Illinois: During the mid-1980s, looters just across the Ohio River found and then sold dozens of conventional Indian relics from a site known as Slack Farm, for which they were prosecuted and imprisoned under Kentucky's legal provisions for the protection of historical treasures. At the time, Illinois had no such law to prevent the occurrence of a similar incident within its borders until politicians in Springfield unanimously approved legal safeguards like those in Kentucky.

A few weeks before these statutes were set to go into effect after New Year's Day 1990, a powerful explosion that had been set at the mouth of the Olney-area cave thoroughly imploded the entrance, permanently sealing it. The first fifty feet of the cave collapsed, filling completely the thirty-foot-deep stairwell with tons of shattered rock. Having survived intact the ravages of time and earthquakes for nearly two millennia, its steps would never again be walked upon. Worse perhaps, shock effects were so powerful they damaged the entire subterranean necropolis, rendering it dangerous to reenter. The muffled blast in this remote part of Richland County was heard by no one except the person who detonated it.

Immediately following the explosion, a small earth mover, or Bobcat, scraped back and forth over the collapsed entrance with its furrowing blade to hide all traces of the location. Shrubs and large rocks were then hauled across the disturbed ground to further conceal what had been there.

Naturally, Burrows was blamed for the explosion by his critics, who were convinced he set it off not only to thoroughly obscure the whereabouts of the site, but also to prevent anyone else from getting inside it. They accused him of having been expelled from the Civil War reenactment club, in which he was a colonel, when its organizers suspected him of stealing five kegs of gunpowder used to blow up the cave.

Burrows dismissed these accusations as "the reaction of small-minded men," and no one was able to prove his personal participation in the explosion. He could not, however, suggest another likely culprit, and in fact denied it had ever happened. In any case, until January 1990, whoever had access to the cave was free to plunder its contents. Only after Springfield's antiquities protection safeguards became law would such people face severe penalties. The person or persons who so violently closed the site just when they did were obviously aware of pending legislation that would render them liable for prosecution (as it had the Slack Farm looters) if they ever returned to the cave. Rather than share the remainder of its unlooted treasures with the rest of the world, they sealed it off from any future entry, supposedly in order to preserve their monopoly on the sale of its artifacts.

Burrows used to claim that the cave had been "permanently closed" somehow in 1986. This statement was at odds with a photograph taken by Jack Ward of a gold coin apparently removed from the site and marked in Ward's own handwriting, "Found 3 Nov. 1987, No-2 Crypt, Wt. 8.27 oz." A November or December 1989 date for the explosion that closed the cave seems credible. In later years, Burrows amended his previous explanation to say that the site was shut in that year.

After word began to circulate that "Burrows blew up the cave," his reputation dropped to an all-time low. But his critics rarely seemed to be willing to put themselves in his shoes. They forgot that he had not deliberately set out in 1982 to find the site. Its discovery was an accident for which he was totally unprepared. At the time, he had been earning low wages as a woodworker who knew nothing about archaeology outside of an occasional petroglyph he might be able to sell to Thelma McClain for a few dollars. Suddenly, a vast golden treasure was his alone for the taking. Few men, regardless of academic background, would have been able to resist pocketing as much gold as they could lay their hands on. As for the artifacts, university-trained experts to whom he had in all innocence shown the often beautifully made pieces insulted him by implying he had faked them. As it appeared to him, their isolationist bias prevented them from even considering his finds. So, why not sell these strange specimens to people who could at least appreciate and value them?

Beyond fleeting mention in a few obscure amateur archaeology newsletters with at most only several hundred readers, Burrows Cave thereafter slipped into virtual obscurity, and doubtless would have passed entirely out of sight without earning even a footnote in the history of contemporary archaeology. But something took place in the last decade of the twentieth century that was to resurrect the controversy as never before and raise it to popular, even international, recognition.

From its premiere issue in the summer of 1993, *Ancient American* was the first and only magazine devoted entirely to presenting evidence for the arrival of overseas visitors to the Americas hundreds and even thousands of years before Christopher Columbus planted the flag of Spain in the New World. The magazine soon achieved a nationwide readership, and after a few issues had been published, it went on sale at book stores in three Mexican cities and in Brussels, Belgium, with additional readers in Japan. Other publications, even the venerable *Atlantic Monthly,* began referring to *Ancient American* as a unique source for information about the prehistory of our hemisphere. The colorful bimonthly was even avidly (if secretly) read by conventional archaeologists who were supposed to be passionately opposed to the idea of cultural diffusion.

In the summer of 1993, Wayne May asked me if I could help him put together the premiere issue of the magazine and stay on as editor in chief. Because my background was in journalism as a reporter and book reviewer, it seemed I might be able to describe the latest finds and conclusions of cultural diffusionists for a general audience. *Ancient American* was supposed to be a popular science periodical aimed at reaching as wide a general readership as possible, with no pretenses as a scholarly journal for a narrow subscription of academics. As was my publisher, I was an avocational archaeologist, and believed that an ancient people who could build the Great Pyramids or an astronomical computer at Stonehenge were certainly capable of achieving the relatively easier task of building ships that took them to America. But I was unprepared for the story that he wanted to present in our second issue. As mentioned earlier, May had made an important discovery about the local origins of the engraved black stones from Burrows Cave. Now I saw photographs of them for the first time, but was not initially impressed.

The artifacts seemed as newly executed as they were simply rendered. They displayed an unlikely jumble of what appeared to be American Indian, Phoenician, Egyptian, Roman, black African, Hebrew, even Christian symbols, scripts, and attire. Like most other cultural diffusionists, I was fully prepared to accept evidence of the arrival of any one of a number of overseas visitors to our continent in pre-columbian times, but not in such a hodgepodge combination as the Burrows Cave material offered. Five discernibly different syllabaries were represented in the collection: hieroglyphic Egyptian; paleo-Hebrew; North Semitic (Carthaginian or Phoenician); an unrecognizable written language referred to (for lack of any other name) as the Burrows Cave script by investigators; and Ogham, an alphabetic script composed of twenty letters in the form of vertical, horizontal, and angled strokes and used in Ireland and Scotland as late as the fifth century A.D., but with roots in west-central Italy's Etruscan age (circa 500 B.C.), and probably much earlier.

Nothing, however—except a hoax—seemed able to explain how this improbable mishmash of widely divergent races, cultures, and epochs came together in southern Illinois and no other place on Earth. May was less skeptical, and *Ancient American* continued to run articles favorably disposed to the ancient origin of Burrows Cave. Our magazine pulled the controversy from the edge of oblivion, and thrust it before the general public for the first time. May discussed it in national radio broadcasts, such as *The Art Bell Show,* while my articles describing the site were occasionally picked up and republished by other magazines with larger circulations.

We both lectured about the southern Illinois collection at archaeological conferences and symposiums across the country, even though I personally agreed with Dr. Curtiss Hoffman's fifth hypothesis—namely, that the site is a hoax perpetrated by a person or persons having access to genuine Old World cultural materials who introduced these to the site. But I began gradually to reconsider my dismissal of Burrows Cave as a hoax as I was allowed to examine more of its materials. Some, at least, were not crudely made, but rather elegantly fashioned. And the sheer scope of the collection, numbering a staggering seven thousand individual pieces already accounted for, seemed to argue against fraud. Moreover, the stones presented a growing number of esoteric elements that it was difficult to believe Mr. Burrows or any forger would have bothered to include. Quite often, the most arcane symbolism appeared among the least notable items.

For example, a rather humble stone small enough to fit into the palm of my hand featured the image of a ram's head rising from a pyramid, surrounded on three sides by disconnected glyphs of some kind (see fig. 10.1). The little tablet attracted scant attention, and I received it as a gift from Wayne May, who purchased it from Burrows. The few investigators who glanced at its imagery concluded it had something to do with the astrological sign of Aries, the Ram. But their explanation, as obvious as it appeared, did not seem wholly convincing because of the presence of the pyramid, which fit into no zodiacal considerations. In fact, it resembled the Great Pyramid on the Giza plateau, mostly because its angle as depicted on the stone was the same as that incorporated into the Nile Delta's foremost structure.

In a flash of recognition, I realized that the ram's head was not the astrological sign of Aries after all. I remembered that the curve of the animal's horns was used by mathematicians in the ancient world to signify *pi,* or the so-called Golden Section, designating the ratio of the circumference of a circle to its diameter—a religious mathematical convention that was encoded in the architecture of Egypt's Great Pyramid. The ram's head rising out of the pyramid symbolized the Golden Section employed by the architect in its sacred design.

Fig. 10.1. Inscriptions in this stone from the cave seem to include the *pi* ratio incorporated into Egypt's Great Pyramid. Partially surrounding marks may be numerical notation, defining this artifact as a kind of lithic scribble pad or computation tablet for determining *pi.*

As soon as I began to suspect some mathematical significance in the stone's imagery, I consulted the numerals of various ancient cultures, and soon found Old World correspondents to the disjointed figures etched mostly to the right of the ram and pyramid. The little Burrows Cave tablet was covered with a mix of Latin and Egyptian numbers. It was as though someone once used the stone, in its malleable state, to compute mathematical problems based on *pi.* The man or woman inscriber even made his/her thumbprint on the tablet's left side, where it was firmly held during computation. The images that symbolized *pi* were understandably included to invoke spiritual aid in problem solving. It seemed highly unlikely that Russell Burrows or any forger would have spent such effort to fool potential buyers of a relatively minor piece with arcane imagery that only very few "suckers" like myself were able to recognize.

Then, of course, there was all that gold. It actually existed. Abundant photographic and written evidence confirms that it did. And proportionately small amounts were sold. What became of the rest is unsure, but people involved in perpetuating a hoax do not use a reputed ton of gold to no direct purpose or profit. The money Burrows made has been through the sale of inscribed stones, which have nothing to do with gold.

These and similar considerations eventually forced me to go to the top of Hoffman's list of hypotheses and adopt, with many reservations and more than a drop of caution, his first one—namely, that the site is an undisturbed burial of Old World cultural affiliation dating prior to the discovery of the New World by Columbus. While an Old World provenance for the artifacts seemed generally credible, who brought them to southern Illinois and when (to say nothing of the original purpose of the cave itself) appeared less clear.

By the late 1990s, *Ancient American* had become an advocate for the site at a time when Russell Burrows had frustrated and angered many researchers by his steadfast refusal to disclose its whereabouts, sometimes after promising to escort them to the cave. They considered his monopoly of all knowledge about its location proof that he was exploiting the discovery and selling its artifacts as his only (but lucrative) income since 1982. While the cave items in circulation might be intriguing at the very least, they were essentially worthless because all had been removed from the context in which they were found. A single tablet might contain incontestable proof of ancient Old World contact with the New World, but once taken out of its original setting, it lost all credibility and archaeological value.

Cave skeptics particularly blamed personal collectors who supposedly demonstrated their crass amateurishness by financially encouraging Russell Burrows to loot the site of its contents, right down to the last available stone, and in so doing depreciate the entire find by amassing objects rendered valueless through their translocation. William R. McGlone, Philip M. Leonard Jr., and Rollin W. Gillespie, authors of *Ancient American Inscriptions: Plow Marks or History?*, concluded:

> [T]he current debate over Burrows Cave is a prime example of the fallacy of starting in the middle. The argument has been over the validity of a group of portable artifacts from an Illinois cave that reputedly have Old World scripts and icons carved on them, when there is no acceptable evidence about the circumstances of their discovery.

The question of whether they are epigraphically correct or were made anciently by Old World people is not crucial to their value as evidence of pre-columbian contact, unless they were in the ground in Illinois before Columbus. Starting in the middle, with the examination of only the artifacts that have allegedly been removed from the cave, should be replaced by the more basic study of the site itself—the cave—and its reputed remaining archaeological contents. A normal order of investigation would thus be established, including demonstration of existence of the cave and its remaining contents, time of emplacement of the items within the cave, and, finally, the study of the artifacts themselves. Starting in the middle leaves the debate hanging in midair without any foundation of age or context.[3]

David Barron, the respected president of Connecticut's Gungywamp Society (an organization of New England archaeologists), demanded:

> I, for one, continue to press for some credible evidence of the mystery cave's actual existence. Until that site is confirmed, evaluated, radiocarbon-dated, and authenticated by some reliable experts, I will continue to be highly skeptical. Until some trustworthy and objective viewers actually "see" if a cave actually exists, I will not consider the Burrows Cave materials as being valid. I don't take giant leaps of faith without more substantial and forthright data.[4]

He was seconded by Alexander P. MacGregor Jr., a history professor in the department of classics at the University of Illinois (Chicago), who declared in *Ancient American:*

> The archaeology of the cave will have to wait until there is a cave. At least its historical scenario and a selection of its artifacts are available

for scrutiny in the meantime. Everything here belongs to an imaginary universe, in its origin and mode of construction not essentially different from the composite fantasies of science fiction. In sum, the cave is a washout. There is no discussing it as archaeology, since Burrows refuses to divulge its location.[5]

But Professor James Scherz sensibly pointed out:

[O]nce the cave is again entered and verified, the present security will be broken. Who will pay for armed guards to man the cave twenty-four hours a day? How can we be sure that those who enter will not accompany another group of treasure hunters to the site? Those who want the cave to be entered to merely satisfy their own curiosity until qualified officials take it seriously are not those who are charged with worrying about the cave's security and an interesting human hysteria known as gold fever, which results when people see or hear about heavy yellow metal.[6]

For his part, Burrows offered his own reason for refusing to tell anyone about the location of the site. "The preservation law of the state of Illinois is currently the obstacle for viewing the cave," he responded during an *Ancient American* interview in 1994, explaining that upon notification of the existence of an archaeological site, it assumes jurisdiction of the site, as well as ownership of any remains and artifacts excavated. "I would love to work with the state, but in order to do so, I must then watch while the site is destroyed and the remains of whoever those people were are removed from their graves [and] stored in cardboard boxes or drawers in Springfield . . . We are diametrically opposed to this . . . The cave was sealed in 1989, before the current law was enacted, and it has not been entered since, and it will not be."

In the interview, Burrows also referred to a federal law stipulating that any institution receiving federal funding must turn over to American Indian tribes those remains and grave artifacts that belong to them, and that those that cannot be claimed by any tribe must be turned over to the Miami tribe in Oklahoma. This, he stated, must not be allowed to happen.

In his final statements as to why the cave must remain inaccessible, Burrows told the interviewer that the cave was private property, and thus subject to the landowner's ultimate approval of any activity occurring there. He concluded: "I think that the state archaeologist is being very wise in staying out of this, because it is a no-win situation. Whoever forces the issue is going to get burned, because they're going to have the Native American groups all over them. Look at what happened at the Dixon mound and burial museum recently, and you will see what I mean."[7]

Burrows raised some valid points in defense of his determination to preserve the cave's secret location. Dixon Mounds, in west-central Illinois, was the site of an important thirteenth-century mass burial, the human remains of which were exposed in the state-funded museum there until Indian activists succeeded in closing the cemetery after prolonged, heated debate. Parallels with Burrows Cave seemed legitimate. Dr. Cyclone Covey agreed, and referred to the fate of yet another site whose location had been revealed: "If Burrows or the executor, who also knows the cave's locale, divulged it, the contents would suffer the same fate as Spiro Mounds of Oklahoma."

Spiro Mounds were catastrophically looted of thousands of delicately engraved mollusk shells during the 1930s. "We had better be wary of trashing Burrows," he continued, "who brought the thousands of stones and other bountiful information to light and endured disdain for his service—a familiar fate for discoverers."[8]

However, critics argued that the Illinois Law for the Preservation of Cultural Antiquities was not as draconian as Burrows described it. He "gave details of the Illinois law that he sees as an obstacle to excavating the cave," wrote Evan Hanson in *Ancient American*. "While I agree that this law is overly strict, it still is basically reasonable. It allows plenty of room for granting excavation permits to any qualified group of professionals. Any group who could satisfy Russ should be able to satisfy both Illinois and federal laws. And that is the problem! How can we find a group of professionals who can satisfy Russ?"[9]

True, a layman's reading of the statute seems somewhat vague—but it would appear that archaeological items found on private property do belong to the landowner. And while true that any human remains must be reported to local police within twenty-four hours of their discovery, if proved to be one hundred or more years old, they must revert to the ownership of the person who found them. Burrows's comment about the state archaeologist seems odd, for Mark Sary dismissed as worthless the cave specimens personally presented to him for his assessment by Burrows and Ward. As Scherz reminded *Ancient American* readers, "The Office of the State Archaeologist was aware of all this. People there seemed annoyed or amused, and did not even go to the field to investigate."[10]

Fig. 10.2. Beverley Moseley (center), president of Ohio's Midwest Epigraphic Society and photographer of artifacts from the southern Illinois cave. He is flanked by Cyclone Covey, professor of history at Wake Forest University in North Carolina, and Covey's wife, Gloria.

Moreover, the 1990 Illinois law, with its "Rules for the protection, treatment, and inventory of unmarked human burial sites and unregistered graves," allowed for certified professional excavators and their designated assistants (with or without related academic backgrounds) to investigate and photographically document apparent interment areas of possible archaeological worth. Investigators would determine the disposition of the finds and then decide on either transference to Native American tribes if indisputable cultural affinities were established or removal to scientific institutions for further study and possible public display. In other words, Illinois state troopers could not be expected to raid Burrows Cave, haul away handcuffed archaeologists in arrest wagons, or confiscate thousands of stones inscribed with Phoenician writing for delivery to bewildered Miami Indians.

Undeterred by Burrows's stubborn refusal to disclose the cave's whereabouts, a group of amateur archaeologists decided to look for it themselves. The expense and labor they dedicated to their quest indicated the extent to which the controversy had fired the minds of those touched by it. For three years they combed a suspected area of southern Illinois, quizzing local residents, buying up plots of land, and bringing in earth-moving equipment to excavate possible targets. By the end of 1996, they had spent three hundred fifty thousand dollars in the search.

Their business manager, treasurer, and secretary George W. Lodge wrote of the enormous investment in human energy required by the enterprise:

> Required were dedicated field crews in both Florida, our organizational home, and distant Illinois, site of the discovery. Their members worked weekends, holidays and late into evening hours. I have had to deal with police, sheriffs, photographers, geologists, copyright/trademark lawyers, newspaper reporters, and television producers. This project has not been the ambition or inspiration of any single person, but the cooperation of many people working together toward a common goal. To serve the high historical value of our coming discovery, I have a national news network on standby. An internationally recognized public relations firm is involved, as well as officials from the state of Illinois. I plan to direct the marketing team publicizing this discovery and its importance to the world.[11]

Lodge's grand scheme for developing the cave indicated his certainty that its rediscovery would be shortly forthcoming. He promised to "seal and secure the site, allowing authorized personnel and official researchers only; build a replica of the entire crypt as it is explored, and have original artifacts removed for study, returned, or replaced; work with the authorities in all countries affected by decipherments (of the black stone inscriptions); make all research and information obtained in the past fifteen years available to the general public." But eventually, as Burrows Cave continued to elude all the archaeologists' best efforts, hopes began to wane. "The future of our project is still speculative," Lodge commented in an early 1997 issue of *Ancient American,* "and some claim we are chasing a rainbow."[12]

His associate Harry Hubbard had been certain the cave would be rediscovered before the spring of 1997. But now he, too, raged in a letter to the editor:

> I want Mr. Burrows and the world to know that we are not carrying on a vendetta against him. The only thing we want is the truth, the whole truth and nothing but the truth about the site he found in 1982. It does not belong to him. As a national treasure, its true ownership is far broader. For fifteen years, he has prevented the greatest archaeological discovery of all time from being passed on to its rightful owners—humanity. As things now stand, Russell Burrows will not go down in history as the one who made this unparalleled discovery, but as the man who kept it to himself for as long as he could from his fellow Americans.[13]

Burrows's response was terse and unsentimental: "I refuse to reveal where it is located. If this situation cannot be resolved, so be it. I know full well that should I reveal that location, the cave will be looted and the state of Illinois will do nothing about it." Meanwhile, personal collections of the inscribed and illustrated black stones he retrieved from the site continued to grow. Deplorable as private sales of the Burrows Cave relics may have seemed, they at least gave investigators a chance to decode their mysterious inscriptions. Personally, I felt they must contain all the answers; if only they could be translated, the whole story might be revealed. And, as Dr. John White seriously pointed out, "An important test for the Burrows Cave artifacts will occur if the writing cannot be deciphered or turns out to contain foolishness or modern information."[14]

Like others mystified by the enigmatic inscriptions, I tried to make some sense out of them. While mostly unfamiliar with Semitic written languages, I have some acquaintance with Egyptian hieroglyphs, and chose to tackle these specimens from the Burrows Cave hoard. I consulted Budge's translation of the *Book of the Dead,* not a book, actually, but a collection of mortuary texts, spells, or magic formulas placed in tombs as spiritual aids for the souls of the deceased on their journey to the afterlife. I discovered that the same sign emblazoned on so many

Burrows Cave tablets appeared occasionally throughout the *Book of the Dead*. The glyph was pronounced "pet," and signified "heaven." Perhaps its appearance inscribed next to the stone portraits of military men implied that they had died ("gone to heaven") since their likenesses had been made (fig. 10.3). Thus, the "pet" glyph would have served as a kind of pre-Christian crucifix denoting death.

From Beverley Moseley, who owned perhaps the largest visual record of Burrows Cave, I obtained photographs of hieroglyphic inscriptions found on the black stones. Every sign was listed in Budge's Egyptian dictionaries. Curiously, however, all the Illinois hieroglyphs were mirror images of their Egyptian counterparts. It would seem that even an ignorant forger, using the same sources readily available through any public library, would have copied them out correctly. Duplicating them in reverse seemed to undermine their credibility. What did it mean?

Fig. 10.3. One of the cave's portrait stones featuring a soldier accompanied by the Egyptian hieroglyph for "heaven"

Despite their backward appearance, finding the meaning and pronunciation of the glyphs was not difficult. And I was surprised to see how the Burrows Cave inscriptions fell into comprehensible sentences. Moreover, they read properly, top to bottom and right to left, just as they do on temple walls in ancient Egypt. A particularly interesting specimen read, *Nu uten rut ee mer sha ah,* or, "The very beloved sacred dead perished (were cut down) at sea." In view of the hazardous transoceanic voyages apparently undertaken by the ancient inhabitants of Burrows Cave from the Old World to southern Illinois in prehistory, this statement seems as appropriate as it is poignant. Another intriguing cave inscription declared, *Buh per hetch ee senti hru kerh*—"We (or they) go to the very splendid treasure house of gold on established days and nights."

Beginning in spring 2000, events seemed to converge with my translation attempts. Ground-penetrating radar surveys of the suspected site of the cave did indeed indicate substantial concentrations—a "treasure house"—of underground gold. Long before, Chief Brown, the last elected head of the Yuchi, the indigenous tribe native to this part of southern Illinois, told Dr. Joseph Mahan (fig. 10.4), professor of history at the University of Georgia at Columbus, that Indian oral traditions preserved the tribal memory of light-skinned foreigners who "a long time ago" arrived "from over the Big Sea" to bury "a great treasure of gold" in southern Illinois.

After the Yuchi were forced from the region by their enemies, he told Mahan, his people forgot the site's exact location. This story, recorded twenty-seven years before Russell Burrows found his cave, simultaneously tends to establish the credibility of Chief Brown's statement, the site he appears to have described nearly three decades prior to its discovery, and the hieroglyphs defining a "treasure house of gold."

Thus encouraged, I attempted to understand the cave's Semitic inscriptions, with far less success, however, than I found with my Egyptian translations. Even so, a line of North Semitic letters clearly spelled out *Yod-beth-qoph-he,* something to do with "the love of God" *(Yod-beth).* Interestingly, the stone on which this particular inscription appeared included a Phoenician glyph for Tanit, the chief goddess of ancient Carthage. Although I could get no further with the Semitic written statements, I began to understand how they nonetheless, even in fractured form, comprised internal evidence for advocates of the collection's authenticity, although a knowledgeable forger might have reproduced the same lines from books about Phoenicia available at just about any public library.

Fig. 10.4. Dr. Joseph Mahan, who learned of the cave, its whereabouts, and its contents from the Yuchi Indians twenty-seven years before it was found

Seeking an expert opinion, I contacted a linguist at the Oriental Institute of the University of Chicago. I hoped he would at least be willing to examine with an open mind a few examples of the inscrutable

Fig. 10.5. One of the numerous inscribed texts recovered from the cave. Its translation may reveal the Mauretanian drama that unfolded in southern Illinois two thousand years ago. Photograph by Beverley Moseley

script. The professor was a world-class authority specializing in written languages of the ancient Near East, particularly Phoenician, which seemed to dominate the stones. "I have seen several specimens of Old World script supposedly found in America," he said, "but they were all bogus." He was nevertheless willing to look over inscriptions from Burrows Cave, of which he knew nothing, but only on condition of anonymity. Having assured him that his name would not be used without permission, I gave him clear photographs of five different tablets covered with writing (fig. 10.6).

He said they resembled an impossible mix of characters taken from the Phoenician alphabet and jumbled with many other unknown signs in an apparent effort to create a new (unknown?) writing system:

Fig. 10.6. This trilingual inscription from the cave in what appear to be (from left to right) Numidian, Celtiberian Ogham, and North Semitic resisted translation by a scholar from the University of Chicago's Oriental Institute. Photograph by Beverley Moseley

It is, of course, common for a writing system to be adapted for use to write in another language, but not common, to my knowledge, to take an alphabetic system with a small number of signs and use it as a base for a much more extensive system, syllabic, for example.

Also, the disposition of the signs on the objects conforms to nothing West Semitic from the first millennium B.C. such as the longest text with several organizations of signs made up of squares and rectangles. Some of the signs that I recognize clearly belong to different periods and writing traditions from the first millennium B.C., which indicates to me that someone used several script charts to put the fake inscriptions together. They were made by someone who had charts of the old Phoenician-Hebrew-Aramaic scripts. But these easily identifiable signs are so mixed up with others, they make no sense. I see no basis on which to identify these texts as authentic old Semitic inscriptions.[15]

Deleting his identity, I shared these remarks with Cyclone Covey, among the intellectual heavy guns on the side of cultural diffusion. I was interested to learn his reaction to a fellow academic I described to him as "an apparently honest expert." Covey wrote in personal correspondence:

> I have grown suspicious of professed truth seekers who fear jeopardizing their reputation if not anonymous. In one view, this is prudence. In another, cowardice. In either view, truth professing is secondary to something else. Your "honest expert" might also have spoken a little less arrogantly about something he could not explain as outside his expertise. He *assumed* faking, therefore *assumed* recent use of alphabetic charts . . .
>
> The Mediterranean letters looked nonsense to this Semiticist, who seemed to forget Numidian and Celtiberic were also written in Mediterranean letters. Why couldn't he as well have said, "These are not Semitic, but could be Numidian or Celtiberic . . .?" Instead of "many charts," we might conceive many ethnic groups comprising Mauretania.[16]

Numidian and Celtiberic were languages spoken by the inhabitants of what are now Libya and Spain, respectively, during Roman times, and Mauretania, as we know, was an independent kingdom, later a Roman colony, in northwest Africa. Covey believes influences from these sources are evident among the Burrows Cave artifacts. "There is a consistency in all the stones," Covey wrote. "I have studied their inscriptions for thirteen years, and can tell you that any snap judgement is puerile." He fired off a defiant postscript: "You can print this if you wish without withholding my name. I may be wrong, but I refuse to be a coward."[17]

Indeed, unexpected revelations occasionally popped up to underscore the stones' ancient authenticity. While I casually examined a copy of the Numidian alphabet featured in an issue of *Ancient American* magazine, I noted that the bizarre characters seemed somehow familiar. Going to my large photo collection of Burrows Cave artifacts, I found several inscribed with Numidian letters. Investigators have long believed that the lithic portraits represent prehistoric visitors from North Africa, where Numidia, as we now know, was a powerful kingdom in what are today Libya and Nigeria, until conquered by Julius Caesar in the first century B.C. Juba II, king of the people depicted on the stones, was, you will recall, Numidian.

I found it difficult to believe that a man of Russell Burrows's nonacademic background would have been subtle enough to include the heaven glyph in manufacturing forged artifacts, or known anything about the obscure Numidian written language. Skeptics argued that if not Burrows, then an accomplice: Perhaps, for example, his schoolteacher brother-in-law, Sam Eyer, was the forger. But not even the merest evidence suggested such a possibility. I began to be convinced that the cave inscriptions, however much condemned by most epigraphy experts, were authentic and had been inscribed by overseas visitors from the ancient world to Illinois many centuries before Columbus.

While I had reached my own conclusions about the stones' inscriptions, others continued to debate and ponder, in some instances focusing on the symbol IH/ (see fig. 10.7), which was a secret sign adopted by early Christians who felt themselves too persecuted for the public use of crucifixes. For Burrows Cave researchers in the 1990s, its relatively infrequent appearance among the vast collection was particularly disturbing. How could they reconcile the occurrence of this Christian sign among so many other incongruous, even antithetical symbols? To suggest that in coming to America, early Christians somehow made common cause with pagan Romans, Phoenicians, Egyptians, West African blacks, and Hebrews seemed totally beyond belief.

But again, the proliferation of so much arcane imagery made even the most bitter skeptics pause to reconsider the stones' ancient authenticity, difficult as they found it to believe that their discoverer and salesman was capable of forging such archaeological esoterica. Yet, it was precisely because of their "internal evidence" of pre-columbian visitors that the inscribed stones continued to fetch handsome prices for nearly twenty years. While investigators were torn over these annoying inconsistencies, the most serious effort to enter the cave was under way.

Fig. 10.7. In this inscribed stone from the cave, the Hebrew sign for Yahweh is clearly defined at the top. At center is an apparent representation of the cave itself. The written language appears to belong to an obscure Semitic tongue. Photograph by Wayne May

11
Where Is the Cave?

Reconstructing ancient cultures by looking at potsherds and other artifacts is comparable to trying to deduce the plot of a stage play by examining the props.
JOHN TIFFANY, EDITOR, THE BARNES REVIEW

In 1999, Ralph Wolak, a documentary filmmaker from John King Productions, in Escondido, California, persuaded Russell Burrows to open the cave for his camera crew. In an agreement signed by both men, Burrows promised to take them to the site, where he would be paid five thousand dollars cash. After agreeing to this, he was obliged to identify the location of the entrance to the subterranean rooms. Once the investigators could see that such an entry did exist, he would receive from them an additional twenty thousand dollars, also in cash, regardless of what might or might not be found inside. Their agreement specified that Burrows was required to remain with the team until the entrance was uncovered. A university-trained geologist and Dr. James E. Gillihan, a professional archaeologist from Wabash Valley College, in Illinois, were on call. As soon as the cave was opened, they were to personally examine its contents and interior before anyone else was allowed inside.

Wolak's associates included *Ancient American* publisher Wayne May, three diggers armed with picks and shovels, and Gary Mitchell, the owner/operator of a particularly sensitive metal detector. The instrument was specifically designed to detect artificially created gold objects under the surface of the earth—not unmined gold, but metal that had been fashioned into man-made artifacts. The cost of operating this sophisticated tool was fifteen hundred dollars per day. Burrows insisted he had removed only a fraction of the gold that still remained in the cave.

On March 24 he led May, Wolak, and their associates to a remote location twenty-five miles northwest of Olney, just south of the Embarras River, across a big field and up a hill overlooking the river only a short walk away. They noticed almost at once that part of the hillside had subsided, apparently in the recent past. The feature resembled a decided dent, as though an internal explosion of some kind had been set off to create a slightly collapsed depression. They observed, too, that a road one hundred feet from the fence of the property belonging to the adjoining farm offered easy access to the location—not far for someone carrying artifacts from the cave to a pickup truck or the trunk of a waiting car.

May found large numbers of old pottery shards scattered throughout the vicinity, plus the faint though telltale remains in the earth of either three or five prehistoric mounds. All these details encouraged the men in their hope that Russell Burrows had indeed brought them to the secret tomb. He allowed himself to be photographed for the summer issue of *Ancient American*—standing atop the hill, he pointed down at the ground and was quoted as declaring, "This is the Cave!" Apparently, all his deep concern about the Illinois State Law for the Preservation of Cultural Antiquities had been forgotten. He did not notify the state authorities now, and gone was his fear that he would have to "watch while the site is destroyed and the remains of whoever those people were are removed from their graves" and "stored in cardboard boxes or drawers in Springfield."[1]

Mitchell began sweeping the spot with his metal detector. Almost at once, its indicators lit up, suggesting that a massive accumulation of precious metal items lay somewhere beneath where they stood. Burrows was handed five thousand dollars in a bound stack of fifty crisp hundred-dollar bills, which he snapped into a briefcase without a word of thanks. While Wolak's men continued to claw furiously toward the suspected gold, Burrows slipped away to Centralia, where his son was engaged in a football game. Following an hour of unrelenting excavation, all digging came to an abrupt halt when an impenetrable ledge of solid rock was uncovered. It appeared undisturbed, and everyone wondered optimistically if Burrows had miscalculated by several feet in any direction. But with their renewed efforts, it became apparent that they had been deliberately misled and abandoned by the man who had been paid for his assistance. Some wanted to go after him immediately to retrieve the five thousand dollars and to personally express their disappointment to Mr. Burrows in forceful terms. But Mitchell urged calm and advised them to return to their excavations. The gold was definitely there, he insisted. They might be able to find it without Burrows, after all, and save their investors his twenty-thousand-dollar finder's fee.

Alarmed at Burrows's behavior, Wayne May began to wonder if they were really free to dig up the area that Burrows had nervously assured them belonged to his gracious, still anonymous landowner. A local real estate agent May sought out informed him to his horror that the property in question was privately owned by a farmer who had never heard of Russell Burrows, let alone given him or anyone else permission to dig up his field. May profusely apologized for the unauthorized excavation and promised to make restitution for any damages to the property. But the landowner, Donald Bougham, was magnanimous. He gave permission for the out-of-state strangers to go ahead with their digging, as long as they did not disturb his bean field.

Wolak hired a mechanical backhoe, and its operator went to work tearing great gouges out of the suspected hillock beside the rock wall that had earlier frustrated the efforts of his men. Unknown to them at the time, the vanished Burrows was busy telephoning Mr. Bougham, urging the farmer to have May, Wolak, and the rest, whom he described as treasure hunters, arrested for trespassing. Bougham had never heard from Russell before and hung up on him.

Meanwhile, everyone stood watching expectantly as the mounds of soil, sand, and limestone began to rise. But the earthmover revealed no entrance to a subterranean chamber. Shortly thereafter, in conversation with Wayne May, Burrows admitted he intentionally misled Wolak and the others when he proclaimed, "This is the Cave!": "Those guys were nothing more than treasure hunters," he reiterated. "They lied to me, even though they gave me a five thousand dollar 'gift.' So, I lied to them."[2]

The Embarras site was, he admitted, "something," but it was not his cave. Burrows Cave was elsewhere, at a location he alone knew. In any case, the "treasure hunters" had only enough money left to repair the damage their backhhoe had wrought on farmer Bougham's land, plus a few hundred dollars left over for the long ride back to California. They were intent on seeking satisfaction from Burrows at his Colorado home, to which he'd moved after leaving Olney in 1996, but eventually realized, to their added disappointment, that the written agreement they'd signed with him was badly flawed; they had no legal hold on him. Ralph Wolak and company would never see their five grand again.

Even so, the gold detector's unequivocal readings convinced them that the site, although perhaps an alternative location to the cave itself, warranted further investigation. A few weeks later, they returned with hardened steel bits freighted in from the West Coast to the bean field in southern Illinois. A locally leased drilling rig was set up atop the previous dig, and began probing for the entrance. Several test holes burrowed deep into the rock ledge, with no results. But on the fifth try, the drill broke through the roof of a subterranean space.

A cheer of discovery went up from the excavators, but their rejoicing was premature. The drill had dropped into a cavity only two feet deep, far too small for any kind of a burial chamber. Drilling resumed a few yards away. Suddenly, a fountain shot from the ground. Some kind of natural aquifer had been tapped. Hope rose with the gushing geyser, because everyone remembered Russell Burrows's description of a subterranean water source above the rooms of gold and carved stones. But the drilling had to be stopped. If allowed to proceed, it might flood the artifactual chambers located perhaps directly beneath. Continued drilling along the Embarras site failed to reveal an entrance or position of the precious mineral deposits still registering on Mitchell's metal detector.

Funding for the project dried up, and the investigators had to abandon all work. But their efforts were not without some compensation. They made several important contacts in the area and followed their leads to some wholly unexpected discoveries. As though one underground chamber filled with ancient artifacts in Illinois were not enough, Wolak found two more. Although far less spectacular than the purported treasures of Burrows Cave, the pair of subterranean sites magnified the mystery. Both were located about forty miles from Olney, following Route 50 southwest into Marion County. There, just north of the smaller town of Iuka, in another farmer's field, Wolak was shown the hidden entrance of a tunnel dug into the side of a large hill.

The passageway, approximately fifty inches high and six feet wide, was obviously man-made. After running 187 feet, the tunnel made a slanting turn to the left, narrowing to about three feet across and descending another forty feet at a forty-five-degree angle. It finally opened into a roughly square room some sixteen by sixteen feet. Carved out of its left wall (as seen from the entrance) was a raised rectangular platform resembling a funeral bier. In fact, John Scully, the landowner who discovered the underground complex, claimed to have found a human skeleton laid out on the flat projection. He removed all bones from the chamber and buried them somewhere on his property. The skeleton was missing when Wolak arrived. At the same site Mr. Scully also found a magnificent steel halberd (fig. 11.1), a kind of war ax, in virtually pristine condition, lying on the stone bier. The object was later tentatively dated from sometime between the eleventh and fourteenth centuries A.D. by specialists in weaponry of the European Middle Ages. Otherwise, the room was empty.

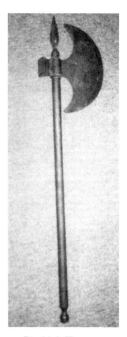

Fig. 11.1. The pre-columbian halberd found in another southern Illinois cave

Not far away, Wolak investigated yet another subterranean chamber, likewise man-made. It had mostly collapsed, however, so his entrance was restricted. He was nonetheless rewarded with recovering from the fallen rock and earth a magnificent metal shield embossed with intricate, masterfully wrought images of men, horses, and mythic characters involved in some kind of drama or pageantry (fig. 11.2).

While neither of the Iuka sites may be a candidate for Burrows Cave, their discovery is hardly less dramatic and significant. Further testing is needed to determine the precise identities and ages of the halberd and shield. If they prove to be medieval, both will constitute persuasive evidence of Europeans in North America during pre-columbian times. In any case, the existence of two more strangely artificial caves containing archaeological materials, only some fifty miles from the Embarras River site, implies a nonnative prehistoric cultural presence in southern Illinois.

Fig. 11.2. A metal shield retrieved from a subterranean chamber in southern Illinois

To some investigators, the ornate shield at least suggested later French origins, perhaps as recent as the eighteenth century, when famous explorers and trappers like Père Marquette and Louis Jolliet were traveling throughout the area. Interestingly, several mud stones removed from Burrows Cave were allegedly covered with writing even Russell Burrows immediately recognized as French. Their fate was a unique loss still keenly felt by investigators, who believe they certainly would have shed important light on the cave enigma. French relations with the Indians were generally cordial, unlike the mutual hostility experienced by the English and Americans and the Indians, largely because the French were more interested in transient fur trading than in settled farming, which meant depriving indigenous peoples of their land. The occurrence of French-language inscriptions in southern Illinois makes sense. Perhaps the Indians, on whose goodwill the French depended, showed the friendly foreigners an old sacred site—namely, the cave, where the Europeans were invited to make their own written contributions, like those composed by the other white-skinned, bearded men who arrived there many centuries before. Unfortunately, we may never know.

Almost a year after the gold-hunting fiasco, Burrows unexpectedly announced in the June 2000 issue of *Ancient American* that he had received a call from one of the men in charge of the property on which Burrows Cave is located:

> He said that a decision to proceed with a professional examination of the cave was about to be made and urged me to return at once to Illinois. There I was to meet with the local authorities, as well as the anthropologist in scientific charge of the site. They needed my participation because no one else knew the precise location of the cave. I arrived in St. Louis early the next morning, then drove to Effingham, Illinois, where I met the site anthropologist. I at once recognized him as the same gentleman who visited the cave with me in the 1980s.[3]

Who this "gentleman" might have been is a mystery. Burrows never mentioned him during our e-mail correspondence, nor does this figure appear in any of the writings of Jack Ward, who was the earliest co-participant in the cave project. Burrows continues:

> I was assured that both he and his university would not conceal anything. If the artifacts and remains originated from somewhere other than America, such a finding would be reported. I trusted him, and we adjourned to the site, where heavy earth-moving equipment was already in place. Ready to go was one of the biggest bulldozers I ever saw. In just three passes made by this monster . . . the cave was revealed. On hand was a crew of about twenty shovelmen, who cleared out the earth under the overhang, as well as the fifteen-foot plug I put in when it was sealed during July 1989.

During September 2000, Burrows made no mention of "a fifteen-foot plug," which seems outrageously unwieldy to have been managed by himself. "From that point," Burrows explained,

> the cave seemed to be in as good a condition as when I was last inside. The anthropologist informed me that they would begin a preliminary study right away; they planned to inventory and photograph everything they found. Eventually, a full report would be made to the public. I could ask for no more than that. Unfortunately, I was not permitted to claim the two thousand artifacts I stored just inside the entrance. On May 15, 2000, Burrows Cave was opened to persons other than myself for the first time since its discovery eighteen years ago.[4]

According to Jack Ward's unpublished papers, however, four men in addition to Burrows visited the inside of the cave: A.R.E. colleague Norman Cullen; Burrows's brother-in-law, Sam Eyer; the landlord, who was still nameless in Ward's account; and Ward himself. In any case, Burrows went on:

> [A]rrangements have been made for thorough study of its interior to be completed by certified experts from a leading university. The names of the scholars and institution involved will be withheld until the authorities themselves determine a proper time for their disclosure. No one other than the professional investigators will be permitted to be present during the cave's examination. Once preliminary examination of the site has been completed, Wayne May, the publisher of *Ancient American* magazine, will be permitted to enter the cave and photograph its contents, to be featured in a forthcoming issue . . . The only thing I can say is, "It's about time!"[5]

While those who knew little or nothing about Russell Burrows read his article with excited anticipation, most readers more familiar with nearly twenty years of frustrated hopes and continual disappointment immediately noticed the glaring inconsistencies in his article. They had heard such promises many times before and shrugged off this one as just the latest example. Time seemed to validate their scorn for his announcement that "Burrows Cave is opened!"[6]

"You'll be inside the cave in the next thirty to ninety days," he promised Wayne May, who good-naturedly thanked Burrows, then waited to see what the next month or so would bring. Nothing, it turned out. Spring passed, and summer morphed into fall without a word from the discoverer. As far as *Ancient American* sources were able to determine, no university team of anthropologists with "shovelmen" supported by earth-moving equipment and bulldozers were hard at work anywhere in southern Illinois. Their major operation might have been conducted under strict military-style secrecy, but by autumn the "opening" appeared to be a nonevent. The phantom university with its "anthropologist in charge of the site" was no less elusive than the whereabouts of the cave itself, and his or her "full report made to the public" never appeared.

In October, Russell Burrows explained what had gone wrong. "The fellow doing the excavation ran into a snag," he said irritably. "The big block of stone, which, some time in the past, had fallen and blocked the passageway, has settled so that there is no way past it . . . I hope that it [the cave] is left alone before someone gets killed. It is far too dangerous . . . I have explained this to everyone, but that fact doesn't seem to register."

Ancient American readers reacted with something stronger than skepticism to Burrows's story of the unnamed university excavation unfortunately aborted by a dangerous boulder.

When I asked him about the future of the cave, Burrows answered with terse finality, "Too much hoopla. I think it is gone for good." So much for "the greatest discovery in the history of archaeology," as Dr. John White, perhaps Burrows's biggest customer, once dubbed it! "How do you want your descendants one hundred years from now to remember you and the cave?" I e-mailed Burrows. "That I tried to be honest and respectful about it," he replied. "That I did not loot it, and that I tried to respect the dead."[7]

Dead is how all prospects for the future of Burrows Cave seemed to be at this point. But Wayne May had not given up on it. He still believed the Embarras site near Mr. Bougham's bean field site merited further investigation. Beginning in early spring 2001, he launched a series of subsurface scans. Merlin Redcloud, a member of the Ho Chunk Nation, helped persuade his fellow tribesmen to bring their Geosystems ground-penetration radar to the Bougham farm site. This remarkable instrument is the underground counterpart of sonar. Its electronic impulses probe beneath the surface of the earth to reveal buried objects and pockets of unfilled space, just as sonar locates fish or submarines under the sea. Redcloud had been following the ups and downs of the Burrows Cave saga with some detachment, until he was shown a startling object purchased in a small southern Illinois museum shop.

Some twenty years before, Robert Harmon, proprietor of the Calhoun establishment, found a quartet of stone oil lamps near the Fox River. They resembled nothing from pioneer days, but were identical to examples of the Roman era. Redcloud, on being presented with one of the lamps, was less excited about its apparent Old World provenance than its engraved symbols, which he immediately recognized. "These are our signs," he exclaimed, "but this is not our object," because Native Americans did not make stone oil lamps (fig. 11.3).

The Ho Chunk's ground-penetration radar survey yielded startling results. Readouts clearly showed a subterranean series of dead-straight corridors connecting separate chambers. Fine-tuning on one of the rooms revealed what appeared to be a bier—a raised platform on top of which lay a sarcophagus or coffin. The electronic images were virtually identical to Burrows's own diagram of Burrows Cave, as he published it in his 1993 book. They went farther, tracing the underground corridors to chambers previously undetected by the investigators. To them, it was all too good to be true. How reliable was the radar? Had Wayne May and the others properly interpreted the readouts?

Fig. 11.3. A stone oil lamp, which the Indians do not claim as theirs, found in the vicinity of Burrows Cave

To be certain, they raised funds throughout the summer for three additional subsurface surveys in which they used a variety of electronic instruments, some more advanced than the Ho Chunk's Geosystems. The different operators were unaware of one another's work at the Bougham farm site. A particularly sophisticated method brought to bear was the RAP Sub-Surface Detection and Imaging System operated by its designer, Russian-born Igor Zuykov. His device had earlier yielded fine details under contract for the State of Ohio's Transportation Department. By registering the natural acoustic field of the earth—the acoustic resonance field induced in rocks in response to external factors—the RAP System was state-of-the-art equipment.

Results generated by these various instruments were unanimously consistent. Each of the different surveys showed the same subterranean features. In 1999, just before receiving his five thousand dollars, Burrows had told the truth when he declared, "This is the Cave!" But how to get in? By all appearances, the underground corridors and chambers were fragile. A random dig might cause the entire complex to collapse. The entrance Burrows used had to be rediscovered. Additional surveys were conducted by Zuykov in search of the original entry point, but without success. He said it was like looking for a needle in a haystack, given the broad acreages that comprised the site and the painstaking pace at which his radar imaging device functioned. May and his colleagues considered the options. They might excavate a pit directly over a section of corridor that seemed firm, then lower themselves down into it. But a mining engineer advised against such a particularly hazardous procedure. A lateral shaft would be safer for investigators and the cave itself, but more expensive.

Many thousands of dollars had already been spent on efforts to locate and access the site. Raising additional funds from May's few supporters, who had already given so much with no hard results to show for their generosity, seemed daunting. As happened often during the quest for Burrows Cave undertaken over nineteen years by so many hopeful searchers, the goal of their exertions seemed just within grasp, only to slip away at the last moment. Burrows affirmed that the location on Mr. Bougham's farm might be genuinely archaeological, but it was not the site, he said; it wasn't his cave. "So be it," Wayne May exclaimed. "From now on, this place will be known as the Tombs of the Embarras!" However, for all the money and energy devoted to its discovery, it still remained closed.

The latest impasse had brought all progress to a halt, when May and Wolak were approached by Sam Eyer, who had helped his brother-in-law take many hundreds of artifacts from the cave in the mid-1980s. The respected schoolteacher had kept silent on the controversy ever since, but now he offered to reveal the precise location of the entrance. He said that Burrows and he had descended the thirty-foot stairwell, then walked straight down a five-hundred-foot-long corridor. They brought along a wheelbarrow in which they piled up treasures from the first four chambers, which were emptied of all items down to the last stick. Eyer was considered useful in getting this pillage up the steep flight of steps. After they'd removed as many objects as was humanly possible, they'd bulldozed the entry to at once conceal the site and seal all access to it.

Eyer indicated the very spot of the former entrance to Wayne May, who ordered an experienced excavator driving his own bulldozer to cautiously clear away the area. Within minutes, the operator stopped to point out that the ground had been previously worked by another digger—just as Eyer had described. Additional digging revealed the scattered stones of an archlike structure. The entrance to the Tombs of the Embarras had apparently been found.

But not a word of the exciting find was whispered outside the small circle of investigators. They needed time to organize the professional excavators, photographers, reporters, security guards, certified archaeologists, and geologists required to make their discovery unimpeachably scientific. The site's highly controversial nature mandated that every aspect of its disclosure leave no room for accusations of misrepresentation or forgery. Perhaps never again would be found such an abundance of physical objects confirming the prehistoric arrival of Old World visitors in America. From late summer into early fall 2001, May carefully lined up his forces for the opening and penetration of the underground mystery. The moment of truth was set for October 25. But, as so often happened in previous attempts to access the elusive cave, these most determined efforts resulted in a mix of disappointment and encouragement.

Three key pieces of equipment were assembled at the site. The first brought into play proved to be decisively important for the immediate course of action. Eugene Savoy, an expert in electronic subsurface investigative instruments, passed over the target area with a state-of-the-art device that rendered the earth virtually transparent down to about fifty feet. His linear ground penetration radar instantly relocated the subterranean passageway, but also revealed something the other equipment had missed. The horizontal corridor did not run straight and level, as previous, less accurate readouts indicated, but angled downward at a declining slope. This news was vital to the success of May's operation because his excavation would have missed the tunnel had his men dug straight into the hillside.

The linear radar revealed other information. "The main tunnel system was surveyed electronically and (coincidentally?) proved to be virtually identical to the map purporting to be Burrows Cave, published in volume 1, number 4 of this magazine," May stated in *Ancient American*. "Our survey showed that the subterranean openings were very similar to those indicated on Mr. Burrows's map of years before." Savoy's linear radar also detected the unmistakable blast effects of a recent demolition, which caused about twenty yards of rock and dirt debris to collapse into the corridor. Stories of gunpowder blasts inside the cave to seal off its entrance seemed verified by the linear radar readouts.

May went on to declare:

> We were surprised yet again by statements from Mr. Burrows to the effect that the underground site we identified is not his, which, he insists, is more than forty miles away. According to him, the undisclosed location is being excavated by an anonymous archaeology team from an unidentified "major" university. All artifacts removed, examined, and photographed by the unnamed professor in charge will be allegedly turned over to Mr. Burrows as his personal property, to dispose with as he sees fit. Of course, we are happy to honor Mr. Burrows's request that the location which has become the focus of

our labors will no longer be referred to as "his" site, and will henceforth be known, not as Burrows Cave, but as the Richland County Site, or the Tombs of the Embarras.[8]

In any case, many tons of imploded material had to be cleared away, so drilling commenced at once, excavating some twenty feet into the hillside. A larger, more powerful digging machine was needed, however, to remove the rest of the fill and finish the job. May and his colleagues had perhaps another three days of work before breaking through to the entrance. But the digging machine operator had been seriously delayed by their prolonged preliminary work, and previous contractual agreements engaged his services at other job sites. Frustrated that they had come so near and still seemed so far, the investigators reluctantly called off their operation, rescheduling the digger for early January.

12
The Pasture of Fools

You shall understand that which perhaps you will scarce think credible, that about three thousand years ago, or somewhat more, the navigation of the world, especially for remote voyages, was greater than at this day.
Sir Francis Bacon, The Advancement of
Learning and the New Atlantis

A children's workbook challenges its young readers to connect what at first seem to be randomly placed dots on a page. By drawing lines from one to another, however, something recognizable begins to emerge until, when all the points are thus linked, a completed image is revealed. The same method may apply to the investigation of historical events. Facts are isolated bits of information, mostly unenlightening in themselves, unless their relation to other documented evidence is made clear. The farther that relationship is extended to additional data, the clearer a particular period becomes. Such a technique is especially useful in trying to understand how incredible parallels between a cave in southern Illinois and events in ancient North Africa could have come about. To arrive at such an understanding, we must "connect the dots," as it were, drawing lines of inference from one piece of evidence to the next. The more connections we are able to trace, the more recognizable the picture presented in the previous chapters will be.

Our first "dot" in this historical puzzle is the fabulous Mauretanian exchequer, the capture of which was the primary aim of two Roman emperors, first Caligula, then Claudius. Their North African invasion had successfully concluded after a prolonged campaign led to the collapse of Mauretania and Numidia. Pursuit of the enemy beyond the provincial border did not make military or political sense, unless some other factor warranted it. After more than seven long months of difficult struggle, the Mauretanian leaders and their treasury had somehow slipped through Rome's grasp, never to be seen again.

But could they really have sailed across the formidable Atlantic Ocean 1,351 years before Christopher Columbus officially discovered the American continent? Conventional archaeologists are sure they could not have, because shipbuilding technology, they argue, did not allow transatlantic voyages until the fifteenth century. Even so, the Phoenicians' maritime achievements are officially acknowledged, if begrudgingly. They guarded their western provinces so carefully that anyone who bragged publically about sailing to one of their mysterious colonies—Antillia—was judged to have betrayed state secrets and suffered capital punishment.

When curious Roman captains sometimes trailed Phoenician merchantmen as they made for the open ocean outside the Straits of Gibraltar, the Punic sailors scuttled their own ships, rather than reveal classified trade routes. During a voyage to the Cassiterides, or the Tin Isles, of southern Britain during the early second century B.C., a Phoenician captain noticed that he was being followed by a Roman freighter. He deliberately ran his ship on the rocks and escaped with his crew, rather than betray the islands' location. Carthage rewarded his action by giving him a new ship and full compensation for the loss of his cargo. Erathosthenes, the Father of Geography, wrote in the first century B.C. that anyone sailing toward Sardinia or Gibraltar from Italy, until Carthage was destroyed by Rome, would be killed by Phoenician sailors.

To reinforce their domination of the Atlantic, the Phoenicians spread tall tales of enormous sea monsters, savage cannibals, treacherous shoals, ship-sucking whirlpools, vanished fleets, devouring storms, southern waters so hot they literally boiled, floating mountains of jagged ice in the north, and so forth. Such intentionally melodramatic stories were no doubt based on actual experiences that had been greatly exaggerated to awe any potential competitors.

As Beatrice Chanler writes:

> The Phoenicians and Carthaginians had visited the islands of the ocean, going beyond the Pillars of Hercules. But it was the policy of these cunning traders to shroud the true position of their islands in mystery so that they might reserve for themselves a monopoly of purple dye in the Mediterranean. Centuries later, a Numidian king was their successor. He would adopt the policy of that race [the Carthaginians] to which he was kin and guard his secret. All of them had in fact obtained from a lichen *(Rocella tinctoria),* which grows in abundance on the rocky seacoast of the Madeira and Canary Islands, and which produces a beautiful purple color employed to dye wool, silk, and other materials.[1]

The Phoenicians, as well as Juba II's Mauretanians, jealously protected from all commercial competitors the locations of these overseas sources for *Rocella tinctoria, Murex trunculus,* and *Murex brandaris,* as well as other luxury goods. Sometime during the fifth century B.C., this jealousy erupted into a trade war with the Etruscans of western Italy, themselves redoubtable mariners. It ended in an agreement that defined specific spheres of influence in the western Mediterranean Sea and Atlantic Ocean. Until the rise of Rome, the world west of Sicily was divided between Carthage and Etruria. But the fear campaign worked. It was this maritime propaganda that was the Phoenicians' most enduring legacy. Centuries after the Phoenicians disappeared from history, their accounts depicting the Atlantic as some kind of liquid hell persisted, and were in fact augmented by the superstitions that shadowed the minds of most men through the Dark Ages. Some three thousand years of seafaring tradition had been lost with the collapse of classical civilization.

The Phoenicians were not the only great sailors of former times, however. Around 215 B.C., Ptolemy IV, the Greek king of Egypt, commissioned the *Alexandris,* after Alexander the Great. The largest known ship in the ancient world, she was no less than 450 feet long from stem to stern, with a fifty-seven-foot beam and an empty draft of six feet. The floating giant stood more than seventy-nine feet high from the waterline to the top of her tallest mast, and required four oars to steer her. Her *thanite* oars (the longest, projecting amidships, with six rowers per oar) were each an incredible fifty-seven feet in length. The mighty ship featured a complement of four hundred officers, ratings, and deckhands, carried 2,850 marines, and required an astounding four thousand rowers. Accommodations for chief officers were luxurious, including hot baths, a full-size kitchen galley, and even a shrine dedicated to Poseidon, the sea god.

The *Alexandris* was no showboat, but rather a powerful battleship. Catapult artillery could hurl shaped, fifty-pound boulders or weighted, flaming sponges soaked with an inextinguishable pitch up to eighteen hundred feet with precision. Although never given the opportunity to serve in military operations, the impressive vessel drew huge crowds of onlookers wherever she docked, much like those the gigantic German dirigibles *Graf Zeppelin* and *Hindenburg* were capable of attracting more than two thousand years later. The arrival of the *Alexandris* in port invariably signaled a festive occasion during which daily routine came to a halt. Two centuries after she was launched, her hull still survived, docked in a Greek harbor.

While the outsized proportions of the *Alexandris* may have been unique, her double-hull design was not. Large catamarans had already been cruising the Aegean, if not elsewhere, from at least 400 B.C. The chief utility of the catamaran was to navigate the heavy swells not of the choppy Mediterranean Sea, but of the ocean. Neither was the *Alexandris* an anomaly. The *Leontophoros* was another leviathan launched by a famous inventor, Demetrius. His warship carried twelve hundred marines and was propelled during battle by sixteen hundred oarsmen.

His son, Gonatas, built an even larger vessel, a *triarmenos,* or three-masted design, which actually saw combat and played an important role in the defeat of Ptolemy II in 246 B.C.

Nearly three centuries before either the *Alexandris* or the *Leontophoros* sailed, a Carthaginian admiral, Hanno, led a fleet of sixty vessels carrying thirty thousand men and women on a colonial expedition along the coast of West Africa with enough provisions for a thirty-five-hundred-mile voyage, according to his surviving *periplus,* or logbook. Each ship grossed one thousand tons, compared with the 180 tons of the tiny *Mayflower.*

During the mid–first century A.D., in the time of Alexander Helios, Roman ships accommodated up to six hundred passengers and crew, while the largest merchant vessels, two hundred-foot-long *frumentariae,* carried four hundred thousand bushels (or twelve hundred tons) of wheat between Mauretania and Italy. *Frumentariae* cargo capacity was about ten times that of the *Santa Maria,* the flagship of Christopher Columbus. Clearly, shipbuilders in the ancient world produced ships capable of making regular transatlantic crossings. Arguments insisting that sailors lacked the technical means to challenge the ocean until higher levels of maritime development were reached in the fifteenth century are simply erroneous. The *Santa Maria* would have appeared as a mere skiff beside Ptolemy IV's 450-foot-long battleship, with its complement of 7,250 officers and men. And the Phoenicians' circumnavigation of the African continent at the behest of Pharaoh Nekau II in 600 B.C. could hardly have been the achievement of a people overawed by the sea.

A theory denying the existence of a maritime technology that allowed men to cross the ocean two thousand or more years ago can no longer (yielding to the temptation of a deplorable pun) hold water. For the ancients to have possessed ships of high caliber—and for so many centuries—without bothering to sail them across the Atlantic is far less credible than the possibility that they made regular voyages to the Americas.

Some anti-diffusionist skeptics also insist that lengthy trips across the sea were not possible by mariners from the ancient world because sailors, who never ventured beyond sight of land, traveled only by day, beaching their ships before nightfall. Such assumptions are contradicted by the evidence. For centuries, Carthaginian voyages after dark were so ordinary that by Roman imperial times the North Star by which they navigated was known as the Phoenician. In Athene's instructions to Telemachus in the *Odyssey,* she mentions casually that he must "sail on through the night," as though after-hours voyages were nothing special during the Late Bronze Age (circa 1250 B.C.).[2] In the *Aeneid,* a Trojan captain sailing through the night "kept his gaze fixed on the stars above."[3] "And meanwhile Aeneas was cutting the channels of the sea at midnight."[4] "And Pallas stayed close to him on his left, asking him now about the stars guiding their course through the night"[5]— all of which argues persuasively for high standards of nighttime navigation possessed by the ancients.

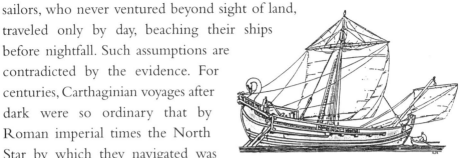

Fig. 12.1. A first-century A.D. Roman freighter (from Dr. Gunnar Thompson's *American Discovery: The Real Story*)

Nor did they lack for necessary navigational instrumentation. In 1900 an unknown device was retrieved from the bottom of the Aegean Sea near the island of Antikythera and was warehoused in the National Museum in Athens. Forty years later, a researcher came across the physically unimpressive artifact by chance and, after a thorough cleaning, discovered it was an analog computer dated to circa 150 B.C. Resembling the intermeshing gears of a complex clock, the instrument provided latitude (not "rediscovered" until the Renaissance), the time of day, and positions of key navigational stars. In the words of the researcher who discovered it, "It was like finding a turbo jet engine in King Tut's tomb."[6] The Antikythera device made possible long-distance voyages beyond sight of land, even across the ocean.

Perhaps someday another fisherman or marine archaeologist will have similar luck finding the Stone of Heracles, a golden cup of liquid with a magnet floating in it, in use from at least the time of Herodotus, twenty-five hundred years ago. Further, in his *History,* Herodotus wrote that the famous Greek mathematician Pythagoras gave Abaras (a magician) a "guiding arrow in order that it may be useful to him in all difficulties in his long journey."[7]

Along a similar line of thought, some twenty centuries before the official invention of the compass by Arabs around 1250 A.D., a Bronze Age compass of some kind is referred to in the *Odyssey,* when Homer mentions an instrument that made it possible for mariners to safely navigate through fog and darkness.[8] The Libyan *tanawa,* or "reckoner," also charted planetary movements and converted from polar to ecliptic coordinates, and may have been similar if not identical to the Antikythera device. Some ships were equipped with it by 400 B.C., at the latest.[9]

These and a very few other navigational devices that have survived, sometimes only in description, demonstrate a sophistication of maritime science not often acknowledged by modern historians. More important, such instruments prove that ancient peoples possessed technologies that enabled them to cross great distances of open water with a reasonable measure of certainty.

A number of researchers and adventurers have set out to prove that transatlantic voyages were possible in pre-columbian times. The surprisingly seaworthy reed boats used so long in Egypt and Sumer were also square-rigged, with planking joined end to end and sophisticated, intersecting framing. These same boats are still sailed along Moroccan shores, where they are known as *madia*. Remarkably, the same reed-boat construction continues unchanged among the Aymara Indians of the Bolivian Andes, at Lake Titicaca. So faithfully have the Aymara perpetuated their ancient inheritance of reed-boat building that Dr. Thor Heyerdahl chose their craftsmanship over the similar but inferior versions still made in North Africa.

Intrigued by cultural similarities of ancient societies separated by sometimes vast oceanic distances, Heyerdahl endeavored to demonstrate in a number of experiments that premodern humans did indeed possess the maritime technology necessary to cross the open seas. His vessel *Ra-I,* constructed in the Arabic Republic of Egypt, began to disintegrate early into his projected transatlantic voyage. But a far more sturdy, seaworthy vessel, *Ra-II,* was built by the Aymara and successfully completed the experiment aimed at proving mariners in the ancient world possessed the means to sail between the continents. Phoenician, Etruscan, and Roman ships were, of course, far superior in every respect to the reed boat Heyerdahl used for his experimental 1969 voyage from North Africa to Mexico.

Even before the *Ra* expeditions, other researchers did much to demonstrate what early man, without the benefit of modern technology, could do at sea. In 1952, Dr. Alain Bombard sailed a fifteen-foot rubber raft with the confrontational name *L'Hérétique* from Casablanca to Barbados, without provisions or fresh water, in sixty-five consecutive days. He carried only a net, knife, and harpoon. The fish he caught supplied all the food and water he needed. The same experiment was successfully repeated twice more six years later by Dr. Hanness Lindemann, first in an African dugout, then in a folding dinghy.

No one hesitates to praise Polynesian sailors for their voyages over several thousands of miles from island to island across the Pacific Ocean in small, open catamarans. Yet, scholars still deny transatlantic capabilities to the seafaring peoples of ancient Europe and the Near East, where material culture was far more advanced than anything ever known in Polynesia. In the *Odyssey,* Menelaus speaks of "a sign on the water of the steady breeze that ships require for *a cruise across the sea*" (my italics), reminiscent of observations made by the far-voyaging Polynesians.[11]

During the early fourth century B.C., the Greek writer Theopompus of Chios composed his *Varia historia Aelianus,* which explained the extent of world geography as it was known in his time. A revealing excerpt reads, "Silenus told Midas about Europe, Asia, and Libya: 'Outwith this world there is a continent or mass of dry land, which in greatness is infinite and immeasurable, and it nourishes and maintains, by virtue of its green meadows and pastures, many great and mighty beasts.'"[12] Silenus added that the continent featured numerous cities of vast and bizarre designs, where inhabitants were governed by a remarkable legal system. Might these lines be describing the American continent? At the time Theopompus set them down, large mesoamerican population centers like Monte Alban flourished in the Oaxaca region of central Mexico.

Further along this line of thought, in his first century *Morals* the Greek historian Plutarch wrote of a continent five days' sail west from the island of Ogygia (Iceland), with a bay, at the same latitude as the Caspian Sea—apparently Hudson Bay. From Ogygia, he instructed, voyagers "are conducted by the winds and transported to the continent of Saturn."[13] Researcher Alban Wall states, "Plutarch set down astonishingly accurate sailing directions on how to reach the American continent from Britain, directions so precise that they provide, in and of themselves, absolute proof that ancient mariners knew full well how to get to these western shores." Wall shows that Plutarch's *Morals* (volume V) describes point for point (V:1, V:1–2, V:3, V:5–6 and V:7) many of the specific features one would encounter on a transatlantic voyage.[14] And the noted archaeologist and Americanist Zelia Nuttall concluded in the early twentieth century, "It seems to me that an accumulation of evidence now forces us to face and thoroughly investigate the possibility that, from remote antiquity, our continent and its inhabitants were known to the seafarers of the Old World, to whose spread of similar forms of cult and civilization in the New World is to be assigned."[15]

Mauretanian ships, hybrids of Roman and especially Phoenician designs, were built for the purpose of long-distance freighting to fulfill the commercial and scientific missions ordered by King Juba II. His numerous visits to the Canary Islands fourteen centuries before their "discovery" by the Portuguese are well documented. He was responsible for naming them after a canine cult of the indigenous Guanches, a neolithic people partially settled on Gran Canaria, which he called Canaria. He christened the island of Fuerteventura with the name Capraria, after the goat-worshiping rituals of its natives. The big island of Tenerife, with its six-thousand-foot volcano, Mount Teide, he named Ninguaria on behalf of its snowy peak. Iuonia, Juno's Island, referred to the queen of heaven, the Canary Island equivalent of Jupiter's wife, revered by the inhabitants of Las Palmas.

The Roman naturalist Pliny the Elder noted in his *Geography* that he received information about the flora and fauna of the Canary Islands from Juba II. The king is credited by Chanler with having gone still farther, reaching Madeira 1,350 years before it was "discovered" by Italian explorers from Genoa:

> This second expedition had sailed from the Purple Islands, traversing a distance of some 625 miles. These navigators showed able seamanship and their knowledge of currents by going straight west 250 miles to avoid a strong east current. Once this space was crossed, the ships found themselves in the zone of the strong north-south currents caused by the trade winds. They could then, with the help of these currents, steer toward the east, certain of being carried sufficiently southward to reach the Canaries.[16]

In the early twelfth century the Arab cartographer Edrisi rediscovered Madeira, writing only that it was heavily forested and apparently uninhabited but featured ruins near the shore. These included a pair of tall stone towers. It seems he did not land, but instead circumnavigated the island, observing the towers from aboard his ship. They may have been built by early Phoenician voyagers, but were just as likely remnants of Juba II's Mauretanian expedition to obtain the *Rocella tinctoria* mentioned by Chanler. When the Portuguese arrived in 1420 A.D. with the intention of colonizing Madeira, they set the entire island on fire. It burned for seven years, destroying any ancient ruins (Punic or Mauretanian) that might have survived the passage of time.

Even more impressive were Phoenician voyages to the Azores, lying 740 miles west of Portugal's Cape Roca. None of the islands was inhabited at the time of their rediscovery in 1430 A.D., but a few important artifacts were found on Santa Maria, where a cave concealed a stone

Fig. 12.2. One of numerous ships depicted on the Burrows Cave stones

altar decorated with serpentine designs, and at Corvo, famous for a small cask of fifth-century B.C. Carthaginian coins found by Portuguese explorers in the early 1750s. By the late twentieth century, conventional scholars regarded the discovery of these coins as apocryphal, until the revelation of an account by Johannes Podolyn, a Swedish visitor to Corvo, in which he describes handling them in 1778.

A more dramatic find was an equestrian statue atop a mountain at San Miguel. The fifteen-foot-tall bronze masterpiece comprised a block pedestal bearing a badly weathered inscription and, atop it, a magnificent horse, its rider stretching forth his right arm pointing out across the sea toward the west. King John V ordered the statue removed to Portugal, but his governor's men botched the job when they accidentally dropped it down the mountainside. Only the rider's head and one arm, together with the horse's head and flank and an impression of the pedestal's inscription, were salvaged and sent on to the king.

These items were preserved in the royal palace, but antiquarians were unable to effect a translation of the "archaic Latin," as they thought the inscription might have been. They were reasonably sure of a single word—*cates*—although they could not determine its significance. If correctly transcribed, it might be related to *cati,* which means, appropriately enough, "go that way" in Quechua, the language spoken by the Incas. Cattigara is the name of a Peruvian city, as indicated on a second-century A.D. Roman map, so a South American connection with the mysterious San Miguel statue might be possible. Cattigara was probably Peru's Cajamarca, a deeply ancient, pre-Incan site. Indeed, the two city names are not dissimilar. In 1755, however, all the artifacts taken from San Miguel were lost during the great earthquake that destroyed 85 percent of Lisbon.[17] The bronze rider's gesture toward the west suggests more distant voyages to the Americas.

Roman accounts of islands nine days' sail from Lusitania (Portugal) describe contemporary sailing time to the Azores. The first-century B.C. Greek geographer Diodorus Siculus reported that the Phoenicians and Etruscans contested each other for control of Atlantic islands that were almost certainly the Azores.[18] With regard to those who might have made the bronze horse, we can recall Corvo's Phoenician coins and that the Etruscans were extraordinary bronzesmiths who favored equestrian themes, such as the example at San Miguel. Further, both the Phoenicians and the Etruscans were outstanding seafarers. But the horse was used by the Carthaginians to symbolize their city, and they also fashioned the figureheads of their ships into horses' heads, so a Punic identity for San Miguel's equestrian statue is favored. Mauretanian origins cannot be ruled out, however. Juba put to good use the seafaring legacy he received from his Phoenician forefathers. "He had inherited a library of Punic books, those of his grandfather, Hiempsal, and, it was said, those libraries of Carthage which the Roman senate had formerly abandoned to the princes of his family," according to Chanler.[19] Among these invaluable documents were no doubt entrusted to him the Phoenicians' confidential sailing instructions enabling the Mauretanians to find their faraway sources of luxury goods, such as the *Rocella tinctoria* dye of Madeira. If the concealed directions told Juba how to reach a small island in the vast mid-Atlantic, perhaps they went further to reveal Plato's "opposite continent," the Romans' vague notion of an *epeiros occidentalis*, on the other side of that "pasture of fools." Juba II would certainly have kept the secret, passing it on to his doomed son, who would have left it as a sole but invaluable legacy to his people—a means for their escape from destruction.

That the Phoenicians not only knew of the Americas but also regularly visited them is confirmed by an abundance of physical evidence. According to American archaeologist Robert Marx, a fourteenth-century Spanish map he found while doing research at Madrid's National Library bears a notation to the effect that a Phoenician expedition launched by Israel's King Solomon around 900 b.c. discovered a land across the sea called Bracir, a Punic version, perhaps, of Brazil.[20]

The map's notation is underscored by a stone, found by Brazilian slaves in Parahyba Province in 1886, inscribed with a Phoenician text. It recounts the two-year voyage from the Red Sea of a dozen men and thirteen women in ten ships. As their fleet rounded Africa, the account relates, they were separated by storms, but managed to make a safe landing on the shores of Brazil, "in the nineteenth year of Hiram, our mighty king" (536 B.C.). Conventional scholars condemned the Parahyba text as a fraud immediately after its publication. "However," Dr. Gunnar Thompson writes, "subsequent research confirms that the inscription is authentic. Passages that were once thought to be erroneous have been verified from identical inscriptions on bona fide Phoenician artifacts. The inscriptions included ancient inscriptions and grammatical forms that were unknown during the 1800s."[21] Dr. Cyrus Gordon, the renowned Semitic-language expert mentioned in chapter 9, was responsible for authenticating the Brazilian find during the 1970s.

Long before the discovery of the stone inscription, in 1641, Jesuit priests found near Brazil's Minas Gerais several bronze figurines inscribed with an unreadable written language. The artifacts were shipped to the Vatican, where antiquarians determined that the statuettes were inscribed in Punic. In 1754, missionaries uncovered additional Phoenician inscriptions on toppled stone pillars in the Amazon jungle, according to Dr. Thompson.

Remarkable evidence for regular Phoenician voyages to Brazil was also found in 1996 by Dr. Mark McMenamin, professor of geology and paleontology at Mount Holyoke College in Massachusetts. Together with his wife, Diana, who holds a master's degree in geology, he previously authored *Hypersea: Life on Land,* selected by the editors of *Discovery* magazine as among the "seven ideas that could change the world."[22] At the University of Massachusetts Library, McMenamin found several published photographs of Carthaginian coins showing tiny markings. They were well known to numismatists, who had dismissed the small amorphous features as errors in production. But the markings attracted his attention because they seemed vaguely familiar. Only after he had enlarged them on his computer did he recognize them for what they really were: elements of a map compiled by the second-century Greek geographer Ptolemy. Through further close scrutiny of the fourteen hitherto neglected specimens of a particular type of Carthaginian coin, McMenamin was able to identify the entire Mediterranean area, including the Straits of Gibraltar, Sardinia, Mauretania, and southern Europe, all depicted on the coins. He even found the Indian subcontinent clearly indicated. In other words, the Carthaginians included at the bottom of their coins tiny maps of locations where they had important ports of call. They were showing off their far-flung contacts on the face of their own currency, just as multinational corporations today boast of their international influence in commercial advertising.

McMenamin was even more amazed to see the east coast of South America represented on the same coins. The outline of Brazil, where so many Phoenician artifacts have already been found, is clearly depicted. His discovery is physical proof, in the form of unmistakable maps, that the Carthaginians came to America not as shipwrecked castaways, but as transoceanic merchants who established regular trade routes between the ancient Old World and Brazil. Reaction to his findings from professional skeptics was predictably hostile, "much of it quite harsh at first," he said.[23] But the configuration of the coin maps, together with McMenamin's impressive academic background, has forced even some of his critics to moderate their opposition.

They have more than his fourteen coins to account for, however. Examples of minted Phoenician currency have been found, according to Dr. Gunnar Thompson, in Alabama, Arkansas, the Bahamas, Connecticut, and Pennsylvania. He has also reproduced Punic inscriptions from Georgia, Massachusetts (Cape Cod), and Nevada.[24] As Dr. Norman Totten writes, "Carthaginian and Roman coins and inscriptions have been found in scattered sites in the Americas. Maritime refugees from wars lost to Rome, especially the final one (the Second Punic War in 146 B.C.), fled to a refuge unknown to Rome—some to America?"[25]

Other interesting evidence has turned up as well. A Phoenician urn dated to Hanno's fifth-century B.C. flight down the West African coast was excavated at Binghamton, New York, and a preponderance of the inscribed tablets removed from Burrows Cave in southern Illinois seem to be Phoenician. Even bones exist suggesting an ancient European presence in the Americas: Spaniels bred by the Carthaginians and known as *techichi* throughout the Mediterranean were observed by Fernandez de Oviedo, a sixteenth-century historian. He noted that the dogs lived along Mexico's Atlantic coast. His report was verified four hundred years later when the bones of a techichi were excavated at a pre-columbian site on the island of Cozumel, just off the Yucatán.

While skeptics continue to insist such information must all somehow be part of a hoax, evidence of Phoenicians in the Americas has continued to surface since Spanish explorers set foot in the New World more than five centuries ago.

13
The Testimony of the Past

> *When we want wheat, but instead get a mouthful of mostly chaff, our natural reaction is to discard the entire lot. But that may be like throwing the baby out with the bathwater. In a scientific age, it should be possible to separate the gems of truth from the matrix of useless rock which comprises the majority of ore in which gems are found.*
>
> JAMES P. SCHERZ, ROCK ART PIECES FROM
> BURROWS CAVE IN SOUTHERN ILLINOIS

The North African refugees had the ships as well as the charts necessary to bring them to America. As Dr. Thompson points out, West Africa "was situated in the most ideal position for launching expeditionary voyages across the Atlantic. The shortest distance across the ocean lies between Africa's west coast at Cape Verde and the Caribbean. Africans and Phoenicians knew from their many voyages that a great equatorial current passed alongside Ghana's coast and headed due west into the Atlantic Ocean."[1]

Although their Carthaginian maps warned them to avoid Florida and Mexico, some Mauretanians appear to have landed along these forbidden shores, driven there perhaps by storms that partially dispersed their fleet. Evidence of this may be found in several examples of native ancient sculpture from the area of Vera Cruz, on Mexico's Atlantic coast. They portray identifiably negroid facial features that seem to have been altered by the patterns of cicatrization that result from the West African practice of permanently scarring the human face for personal beautification and tribal identification.

In 1905, Alphonse Quatrefages, anthropology professor at the Paris Museum of Natural History, wrote of pockets of "indigenous negro tribes" in America, such as Brazil's Charruas, the Black Caribbees of Saint Vincent, and the Jamassi in Florida. Of the anthropology professor's conclusion Gunnar Thompson writes: "Quatrefages believed that these tribes were descendants of ancient transatlantic voyagers from Africa."[2] Might they have been the descendants of the black deckhands and sailors who traveled with escaping Mauretanians from West Africa?

Thompson continues:

> Several anthropologists have identified negroid bones in ancient American burials. A Polish forensic scientist, Andrzej Wiercinski, identified African skeletons at three Mexican archaeological sites: Tlatilco, Cerro de las Mesas, and Monte Alban. A key piece of evidence confirming the West African role in Central American goldworking is the name of the Mexican copper-gold alloy—*guanin*. This term was derived from Ghana. Spanish metallurgists revealed that West African and Mexican alloys contained the same ratios of gold, silver, and copper. Identical alloys are evidence of common smelting traditions which were handed down from master to apprentice. Caribbean natives who met Columbus called the alloy *goanines,* and they told him it was introduced by "black people." The same alloy is called *guanines* in West Africa.[3]

And other physical evidence of a pre-columbian expedition to American has recently been found. In 1988, an American tourist, Archie Eschborn, was scuba diving near Roratan, an island off the Atlantic coast of Honduras, when he accidentally found a Roman amphora half covered by sand. Just sixteen years earlier, sport divers in Honduran waters discovered large numbers of Roman jars, semingly the cargo of a sunken wooden ship long ago dissolved by currents and devoured by sea worms. Additional amphoras were found in 1976 not far from Rio de Janeiro by diver Roberto Teixeira. According to Dr. Thompson, "Scholars identified the amphoras as coming from North African ports."[4]

Another scholar, Elizabeth Hill, a professor of classical Greek history at the University of Massachusetts, was asked to examine wine or olive oil vessels brought to the surface by archaeologist Robert Marx from an ancient wreck near the Brazilian coast in the same year Teixeira made his discovery. After her analysis, she concluded that the amphoras were Mauretanian. Marx also picked up a Roman *fibula*, or bronze clasp used to fasten clothing, in Brazil's Guanabarra Bay, and reported that a stone ballast retrieved from the seabed at Jamaica's Kingston Bay was covered with a Latin inscription.[5]

Fig. 13.1. Romans near Brazil (from Dr. Gunnar Thompson's *American Discovery: The Real Story*)

And perhaps the black African deckhands were not the only members of the expedition from Mauretania to have left their mark. During the late 1890s, explorer Marcel Homet encountered a small, hitherto unknown tribe of blue-eyed, fair-complected, light-haired "natives" deep in the Amazonian interior. They referred to themselves as the Syriana, the Old World name for Syrians. The Brazilian Syriana had been observed as early as the sixteenth century by the infamous conquistador Francisco Pizzaro, who described them as "corn blond." The director of Brazil's National Indian Foundation, Helio Rocha, reported as late as 1975 that the Syriana have inhabited the Amazon since "remote times."[6] An adjunct of the Mauretanian fleet was the Classis Syriaca, composed entirely of Syrian sailors.

Perhaps the largest contingent of ships separated from the main body of the Mauretanian fleet went farthest to the south, landing along the shores of Panama. Living conditions there were probably made impossible by native tribes, who likely threatened the newcomers with incessant attacks. The Mauretanians, glad to have brought with them so many men-in-arms, probably fought their way across the narrow isthmus. Once on the Pacific side, they could have constructed a makeshift flotilla in which they sailed down the South American coast, finally settling in the area of southern Ecuador or northern Peru between what is now Cuenca and Trujillo.

Provocative evidence of their arrival was brought to light by history and anthropology professor Norman Totten. His expertise in the ancient cultures of Morocco and Peru, where he conducted archaeological tours during the 1980s, made him particularly qualified to discern a relationship between several pre-Incan objects and their Mauretanian origins.

In 1978, Totten acquired a pre-columbian Peruvian effigy vessel from Bernheimer's Antiques, in Cambridge, Massachusetts. Previously, it had been part of a private collection going back more than thirty years. Fourteen centimeters high, eleven centimeters wide at the top, and seven centimeters wide at the base, the monochromatic black artifact represented a man's head wearing a turban. This Moche portrait pot, as it was called, was dated to the early first century A.D. The name Moche has been applied by archaeologists to an otherwise nameless people who created a rich civilization along Peru's northern coast from 200 B.C. to 600 A.D. They were skilled metallurgists, pyramid builders, city planners, and potters.

Totten was intrigued by the fired-clay effigy's mustache and inscription; native Indian males are unable to grow facial hair, and the Moche were allegedly illiterate. Epigrapher Barry Fell examined the vessel and determined that the inscription was composed of corrupted Greek letters, as though someone had imperfectly learned the written form of that language. Fell transliterated them to read *Jub Basileus,* or King Juba. Greek was the official tongue spoken in the Mauretanian court. "The letters are degenerate compared with Greek writing in North Africa during Juba II's reign," Totten explains. "Rulers' names and/or titles were frequently abbreviated on ancient Greek and Roman coinages."[7] *Jub Basileus* appears twice on the vessel.

According to Totten, "Doubling designs on pottery was common practice in both Moche and Tiwanakan ceramic traditions." He examined another example of Moche portrait pottery to find that it depicted Berber-like headgear and, most remarkably, a three-lined chin tattoo identical to those still sported by Berber men and women. Moreover, he noticed that the face depicted on the Moche pot, the shape of the eyes, nose, and mouth, "are indistinguishable from Berber."[8] Totten was seconded by Gunnar Thompson, who concluded independently, "Peruvian artists portrayed Phoenician visitors in their ceramics and stone sculptures. The similarities are particularly striking between Peruvians of the Moche culture and Tunisians."[9]

His observations led Totten to deduce that these "Juba pots" suggested the arrival on the eastern shores of Panama of Mauretanian refugees, who then crossed the isthmus to coastal Peru. They did not create Moche civilization, he adds, but "may have been a significant component, together with other components," that left a recognizable impact at least in the shape of portrait vessels.[10] The Mauretanian provenance of his Bernheimer ceramic with its representation of King Juba is underscored by the dating of the pot to the first century, the same time period in which the transatlantic North African expedition took place. The name's execution in corrupt Greek may imply that some semiliterate crew member among the immigrants was responsible for incising the pot's inscription.

Totten's Juba pots are joined by other physical evidence for ancient Mauretanians in South America. Beginning in the 1930s, Ecuadoran Indians inhabiting the Andes Mountains around Cuenca began bringing strange items to their best friend, Father Carlos Crespi, a Catholic missionary from Italy. Following graduation from the University of Milan, he had become a priest and was assigned to Ecuador. In gratitude for his kindness, piety, and medical assistance, he was presented with what would eventually amount to hundreds of peculiar objects the natives claimed to have retrieved from caves deep in the jungle, which were known only to them.

While many if not most of these "artifacts" were hammered together and crudely manufactured by his indigenous parishioners, at least a few showed expertise and Old World mythical themes unknown to them. But the Indians also gave Father Crespi sheets of hammered gold embossed with inexplicable designs and exquisitely wrought solid silver statuettes of Near Eastern deities (fig. 13.2). These were warehoused at his mission, but did not come to the attention of the outside world until the 1970s. Even so, Father Crespi had few visitors beyond his remote congregation, and he never tried to interest archaeological authorities in his collection.

When Father Crespi passed away in 1989, the gold and silver items were partially confiscated by agents of the Vatican, who tried to smuggle them out of the country, but they were apprehended by customs authorities before all the artifacts could be removed. Ecuadoran government authorities seized what objects remained, but refuse to allow public access to them. Still more items are kept at Father Crespi's mission, where they too are under lock and key. Little else is known about

Fig. 13.2. One of several sheets of beaten gold embossed with complex designs presented to Father Crespi by Ecuadoran natives. A visitor from the United States, J. G. Barton, wears a pontifical gold crown, another of the Ecuadoran artifacts. Photograph republished by permission, *Ancient American* 6, no. 4 (May/June 1997)

these controversial objects, because professional archaeologists have ignored the lot, labeling them fakes. While primitive examples among the collections are obvious enough, the massive amount of gold and silver, some of it fashioned into beautiful artwork, seems less easily dismissed. Moreover, links to North African refugees are apparent in some of the Syrian motifs in a few of the artifacts, and in one object in particular: the debated elephant stone. As discussed in chapter 9, an almost exact duplicate was taken from the cave in southern Illinois.

Two outstanding pieces of the Crespi collection are a pair of solid silver statuettes, both about the same size (approximately five inches in height by five inches wide), of a man-headed bull and a winged, eagle-headed man (see fig. 13.3). While these images originated in Babylon during the third millennium B.C., they were adopted by later Assyrian conquerors and survived among the Syrians into late classical times. Both figures represent minor deities, but their identification is singularly appropriate for Syrians who might have traveled with their Mauretanian masters until storms separated them and the former were shipwrecked in Brazil (recall Homet's Amazonian Syriana) and Ecuador.

The man-headed bull is a sphinx, a guardian spirit. In the Crespi version he is portrayed with five legs, which might seem an apparent error attributable to someone who was trying to perpetuate a hoax. But its extra leg helps, in fact, to establish the figure's authenticity: This spirit was commonly, although not invariably, depicted in Syrian temple art as five-legged. Unless a forger were familiar with Assyrian mythology, he would never have created a figure with five legs.

The other statuette is hardly less revealing. It represents a winged genie who dispensed blessings on mortals and gods alike. In his left hand he holds an acorn, containing precious sap from the Tree of Life, shown behind him, to be mixed with the liquid in his right-hand container, thereby creating a kind of holy water, which he sprinkles on those who request benediction. His eagle's head signifies spiritual power associated with the Hindu crown chakra of enlightenment, a spiritual concept appreciated in various cultures around the world.

Both the man-headed sphinx and the eagle-headed genie would

Fig. 13.3. The solid silver representation of an Assyrian guardian spirit (left) and the statuette of an Assyrian genie (right), both from the Crespi Collection

Fig. 13.4. A gold plate from the Crespi Collection apparently portraying an ancient Old World high priest or ruler. Photograph by Wayne May

have been particularly revered by Syrians thrown into a new land, where they were very much in need of protection and blessings from their gods. Due to their similar size and complementary mythic functions, the silver statuettes appear to have originally belonged to a set of religious objects that may have been part of an altar, tabernacle, or shrine. In any case, they are appropriate components of the first-century drama that brought the Syrians, remnants of King Juba II's Classis Syriaca, across the sea to the jungles of Ecuador.

Father Crespi's collection may

be part of the Mauretanian dispersal more clearly defined among the Moche culture, across the border in northern Peru. Together, they tell part of the story of transoceanic immigrants who, separated from their countrymen by powerful storms or hurricanes, finally settled on the Pacific coast in South America nearly two thousand years ago.

14
A Rooster Speaks

When facts challenge theory, theory should be altered to accommodate facts.
Norman Totten, "King Juba Remembered: A Working Hypothesis"

The diverse peoples who were driven to attempt a perilous transatlantic escape from Roman aggression would have included multilingual Mauretanians (speaking Greek in court, Latin in law and civil administration, Egyptian in their temples, and Mauri, Numidian, and Phoenician in military affairs and common speech), who were mostly soldiers; their Iberian and Syrian allies; some West African deckhands; Jews; a few adherents of extremist cults; and still fewer followers of the "risen Christ." It is possible, then, that at no other time in the history of the ancient world had such a wild mix of different races, cultures, and religions been forced together by such a unique set of circumstances. Remarkably, all of these groups are graphically represented throughout the seven thousand or more engraved stones retrieved from a cave in southern Illinois.

For purposes of clarification, they may be classified into a number of general categories: Peoples depicted are Mauretanians, Iberians, Syrians, Phoenicians, Egyptians, Jews, West Africans, and Native American Indians. Written languages represented include Mauri, Numidian, Phoenician, Iberic, Celto-Spanish Ogham, hieroglyphic Egyptian, and paleo-Hebrew. There are a number of "map stones," all depicting what seem to be travel times (with dots signifying days) from the Gulf of Mexico to the cave site in southern Illinois and beyond to areas throughout the American South and Midwest. Another class of these stones is devoted entirely to scenes illustrating the lost myths of obscure mystery cults, but also features better known imagery from Carthaginian, Egyptian, Gnostic, and Christian theology. These are the collection's key elements, which potentially reveal more about their makers than the thousands of more mundane objects in the collection of artifacts, such as simple models of whales and unmarked stone ax heads.

Even these less attractive specimens are not without value, however, and help to establish the collection's pre-columbian authenticity. James Scherz personally examined many of them while cataloging all available pieces, 90 percemt of which have been scattered or lost through sales to private buyers. He observed:

> Rock hammers and axes are sometimes polished to a high degree, but for utilitarian purposes the material must be hard, igneous rock. Limestone of the type that these rocks are made from is too soft for any hard usage. And the extreme shine on some of the pieces and the effigy marks and writing carved into the white hammers speak more of ceremonial purpose than day-to-day usage. And the styles of the points are a conglomeration of the different ancient types found in the fields of the region—from fluted points to chipped points and ground knives. It would seem that rather than being a collection of utilitarian tools from any one period of time that this collection is more like what one would expect from a . . . ceremony attempting to portray in drama the histories and myths of a culture, and provide a plausible explanation to the viewers for the otherwise unexplainable objects in the world around them.
>
> As I handled the pieces, I likened myself to looking at a

collection of props that one would find in the storehouse of a Wagner opera, where the spears and hammers of the Nordic heroes are still portrayed in their ancient form, but are made from material completely unsuitable for the famed weapons of the original heroes. Or I could imagine myself coming upon the storeroom for the ceremonies of a Masonic lodge, where the episodes of the ancient past are acted out by people who are trusted with the different parts of the drama—people who would take great care of their trusted part of the ceremonies, and would likely care for and polish their props.[1]

Among the portrait stones, the Mauretanians comprise the greatest number of depicted faces, and are shown in right-facing profile. Some priests and statesmen are represented among them, and a few images are female, demonstrating that women accompanied Mauretanian men on their voyage. (See figs. 14.1–14.5.)

Most of the heads belong to military men. Appropriately, the Mauretanians were renowned horsemen. They wear a variety of Roman-like helmets, but closer inspection reveals that the headgear is unlike any-

Fig. 14.1. On this white marble block from the Burrows Cave (left), the distinctly non-Egyptian profile of a Mauretanian high priest of Isis, his hair fashioned in the Egyptian manner, appears above an inscription of Numidian and Egyptian characters, including the Egyptian hieroglyph for "heaven." The Burrows Cave stone at right shows a similar image, a priestess of Isis. Photographs by Wayne May

 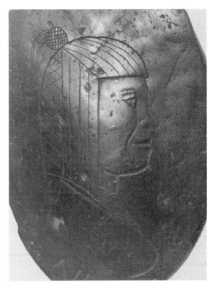

Fig. 14.2. The profile of a woman wearing her hair in first-century Roman style, as she appears on a stone from the Burrows Cave collection. Photograph by Beverley Moseley

Fig. 14.3. The woman portrayed on this stone may have been a person of some importance because she is identified only by the Alexander Helios glyph at bottom. Perhaps she was a member of the royal household. Photograph by Beverley Moseley

thing worn by the imperial legionnaires. Their style, in fact, resembles nothing known, although certain examples exhibit obvious solar motifs, perhaps because they were worn by men loyal to Alexander Helios (Alexander of the Sun). Some helmets must belong to officers, but a number bearing long ribbons or feathers and sometimes featuring a spike at the crown seem to depict cavalry. (See figs. 14.4–14.6.) Almost no two faces are alike. The only exceptions appear to be a young soldier who is represented on at least three different stones, a commanding officer (who is perhaps Aedemon, military leader of the Mauretanian resistance), and, most intriguing and numerous of all, a Christ-like figure who is sometimes hooded.

All the lithic portraits are highly individualistic, as if specific men posed for their portraits. They are distinctly Western European faces, with an occasionally aquiline feature. Until the Arab conquest in 650 A.D., and particularly throughout classical times, the predominant population of Mauretania was Caucasian.

Fig. 14.4. The images of Roman-like soldiers from Burrows Cave

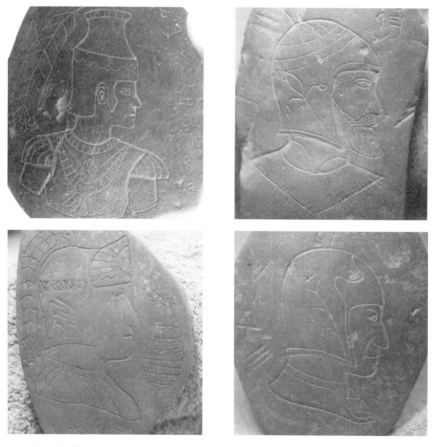

Fig. 14.5. Examples of cavalry soldiers as they appear on the cave's portrait stones

A Rooster Speaks   179

Fig. 14.6. The distinctive profiles of Mauretanian soldiers

The male profiles convincingly resemble the grim faces of veteran combat soldiers—hardened warriors who battled first Romans, then perhaps tribe after tribe of Native Americans as their party made its way up the Mississippi River into Illinois. The inscriptions are almost invariably brief, usually consisting of no more than what is apparently the man's name in Phoenician, together with the Egyptian hieroglyph *pet* for "heaven," implying perhaps that the depicted man was deceased at the time his likeness was completed. Such stones may have been manufactured as memorials to the honored dead. If this reading is correct, it further supports interpretation of the southern Illinois site as a mausoleum.

Though Egyptian was, as mentioned above, the religious vernacular of the Mauretanian court under Cleopatra Selene, high priestess of the cult of Isis, Phoenician was the Mauretanians' everyday and military language, which accounts for the preponderance of its appearance with these portraits (see fig. 14.7). Theodor Mommsen writes:

> The Phoenician language prevailed at that time, so far as there was a civilization in North Africa, from Great Leptis to Tingi, most thoroughly in and around Carthage, but not less in Numidia and Mauretania. To this language of a highly developed although foreign culture certain concessions were made on the change in the system of administration. Perhaps already under Caesar, certainly under Augustus and Tiberius ... the towns of the Roman provinces, such as Great Leptis and Oea, [and] those of the Mauretanian kingdom, [such as] Tingi and Lixus, employed in official use the Phoenician language, even those which, like Tingi, had become Roman burgess-communities. The [Mauretanian] government did not admit ... Phoenician in its intercourse with the communities and individuals in North Africa, but it allowed it for internal intercourse; it was not a third imperial language, but a language of culture recognized in its own sphere.
>
> The heritage of the Phoenician language fell not to Greek, but to Latin. In Caesar's time, Greek and Latin were alike in North African foreign languages, but as the coins of Leptis already show, the latter [was] by far more diffused than the former; Latin was spoken then only by officials, [Roman] soldiers, and Italian merchants.[2]

Mommsen relates that both Juba II and his son Ptolemy were "well read in Phoenician and Syrian literature."[3] In Caesarea, Chanler adds, "Latin, the language of Rome, was rarely heard except for laws and business of state, and even the Punic aristocracy conversed in Greek."[4]

A Rooster Speaks 181

Fig. 14.7. These two portrait stones depicting soldiers are inscribed with Phoenician.

The Mauretanian soldiers' Phoenician connection is underscored by a mysterious symbol that may be seen on at least one of the soldierly stones: three squares placed vertically, each one crossed with an X (see fig. 14.8). Nowhere else is this glyph known to be depicted except on Phoenician burial goods, such as the well-preserved funeral masks buried with the deceased in Carthage to frighten away evil spirits. The X marking may have referred to the common gesture (still used worldwide) of crossing the index fingers of both hands to bar something evil. Although the actual significance of the triple-X sign is not positively known, its appearance on a southern Illinois artifact means that any forger who manufactured it had to be intimately familiar with obscure Phoenician burial practices—which seems unlikely.

In other instances, images of the soldiers are accompanied not by Phoenician, but by Ogham or Iberian script (see fig. 14.9). Ogham was a system made up of notches for five vowels and lines for fifteen consonants, which were etched into stone or the walls of cut tombs to memorialize the dead and/or a visitor. Although the earliest surviving examples of Ogham date only to fourth-century Ireland, connections with runic and Etruscan alphabets bespeak its antiquity, whose ultimate

Fig. 14.8. The triple X—what appears to be a Carthaginian symbol—on this portrait stone of a Mauretanian soldier

Fig. 14.9. A man who may be a Spanish auxiliary on the Mauretanian expedition to North America is identified by a Celtiberian inscription in Ogham.

roots as an elemental script appear to lie in the middle to late Bronze Age. Ogham's identity as an import is evidenced in its signs for *h* and *z,* letters that do not appear in Gaelic.

Ogham's use was widespread from the British Isles to Lusitania, Iberia, and North Africa, with occasional examples found in North America, particularly New England. Appearance of this written language among the southern Illinois objects could be due to its use by Spaniards who preferred it over the more complex Iberian, a preference that seems to typify the military mind-set, which gravitated then, as it does now, toward brevity, clarity, and directness. Already in the time of Juba II, Spain had become Mauretania's favored trading partner, so much so that the cities of Carthago and Gades (today's Cadiz) offered the Mauretanian ruler the honorary magisterial office of duumvir, and Gades additionally granted him its consulship. Many Iberians settled in Juba's kingdom, and some of them could well have joined a refugee ship in its transatlantic adventure. Judging from the Iberian script Barry Fell recognized on the elephant stone from Father Crespi's collection, some Spaniards may have been blown off course, toward Ecuador. The majority, however, appear to have remained with the main body of the fleet, finally landing in southern Illinois, which might be concluded from the relatively numerous cave inscriptions written in Ogham and Iberian.

A French-Canadian specialist in ancient languages, Michel-Gérard Boutet, has concentrated his research on these scripts and their accompanying illustrations, as they appear on the southern Illinois artifacts. According to Cyclone Covey, Boutet is an art teacher (in Laval, Quebec, north of Montreal) who has expertise in Celtiberic Ogham. His Ogham studies were completed in France and Spain. "The Ogham has been particularly baffling on Burrows Cave stones ... Boutet's readings are convincing, because they fit Celtic mythology."[5]

No less baffled than everyone else by the stones' inscrutable written languages, Boutet was startled to discover that the translations of at least some of them became apparent when examined in the context of Celtiberian, the pre-Latin language of Spain: "Being more an iconographer than an epigrapher (my interest in graphic studies led me to linguistics), I began looking into the related iconography of the Celtic cultures and found that everything connected. Words now filled the lines neatly, often spaced exactly where they should."[6]

He was seconded in his conclusions by the noted linguist and Celtic scholar J. Monard, who agreed "that the script was likely Celtiberian."[7] Boutet added:

> Stylistically, however, the Burrows Cave artifacts, as others remarked, seemed strange to the utmost. But then again, looking into the cultures of the proto-Celtic sea peoples, the style compares quite well. The best match is with Punic, Celtiberian, and Celtic art. The problem isn't with Burrows Cave. The problem is with us, our cultural biases and our failure to see them for what they are truly; that is, records on antique visitors to the New World. When Ptolemy II, pharaoh of Egypt, needed to enforce his army, he had four thousand Celtic warriors and their families sent to his service. At that time, Celts were serving as mercenaries in the armies of many foreign kings, those of Syracuse, Syria, Persia, and even Carthage. It is therefore not surprising to find Ptolemaic influence in Celtic art and culture."[8]

Turning his expertise in Celtiberian to the cave's stone inscriptions, Boutet was able to translate a brief example that read, in the Celtic language, *Dru uauda ipu tdau itupe idprd,* or, "The very wise of the mare-goddess-people, therefore also true charmers."[9] His translation is complemented by the cave's numerous gold coins emblazoned with a horse's head. "Mare-goddess-people" is a clear reference to the Celts, who revered the horse as a cultic image.

Boutet's translation indicates that Celtic astrologers (the Unixsuues) celebrated their people's national holiday (Calendar Day B) in a fertility festival connecting with plant life when the constellation of Leo reached a certain position in the night sky. This and all ritual activity or custom had been inscribed on stone tablets presented at a special assembly the time of which was specified by the Celtic calendar.

Its pronunciation in Celtiberian produces a convincing Welsh-like sound: *Le litause didalt cigeri datlh tecon lation uect.* Boutet writes that "this tablet informs us much on the rituals surrounding the depositing of records. It becomes evident that the records were deliberately deposited in a manner respecting customary rite."[10]

Fig. 14.10. The cave's Rooster Stone inscription was the first Burrows Cave stone to yield a translation.

Perhaps Boutet's most impressive translation of the cave's Celtiberian inscriptions belongs to the so-called Rooster Stone (fig. 14.10). Previously among the most mystifying of the cave's numerous images, it features a bird's head with its beak open as though speaking, with Ogham and other written characters in front of its face and at the bottom of the stone. The characters read, *O-Gallebiuo Odete te! Nedsc. Wdah inkuesh data'n,* or, "To the cry of the Gallic rooster, so precious art thou! The cord of ritual binding. Knowledge of Fate-Knower here given."[11]

That these written words should so handily translate into the Celtic language is marvelously underscored by the stone's rooster, which was the national emblem of the Gauls from ancient times, and which the Gauls' descendants still identify with the modern nation of France. It was in Gaul, too, that the Gnostic teacher Marcus actively spread the doctrine and ritual practices of that esoteric religion. The supreme god of the Gnostics was Abraxas, who was depicted as having the head of a rooster (fig. 14.11). In the guise of this animal, which appears to awaken all other life with its cries at dawn, Abraxas was believed likewise to awaken the human soul to its spiritual destiny through the revelation of *gnosis,* or secret knowledge. In Boutet's rendering of the Illinois stone's inscription, "the cord of ritual binding" may be a reflection of Gnosticism's arcane ceremonial practices. His translation "Knowledge of Fate-Knower given here" may parallel one of the surviving Gnostic tenets, "Fate is real, but after it the astrologers are no longer right."[12]

Fig. 14.11. A Gnostic gem from first-century Egypt featuring the image of Abraxas

The most important challenge to investigators like Boutet is to render credibly the meaning of a ubiquitous symbol included on virtually every one of the stones retrieved from the cave. This symbol resembles the Roman numeral for 7, VII, with an inverted V (see figs. 14.12–14.14). Because this glyph sometimes appears in close connection with images of the sun, I asked Covey if it might be the name of Cleopatra Selene's twin brother, Mauretania's last ruler, Alexander Helios. He replied:

> An excellent question. The syllabic letters, as you know, recur redundantly on the stones, often as a caption of a portrait, as if the name or title of the king whom the mausoleum/ceremonial center celebrated. This personage is depicted as a young man, then in stages of aging, in both Roman gear, at least once including Egyptian crown with cobra uraeus, and in Indian garb (which was very similar to Libyan, Getulian, and Garamante). So, he is central, as if founder, honored for centuries after his death.

The burial in the main crypt in a sarcophagus on a slab was consistently Ptolemaic Egyptian. The crypts were identical to those in the Valley of the Kings at Luxor's west bank (a number of which I entered). So, everything fits a Helios hypothesis. Boutet, if I recall right, agreed on the pronunciation of II as "he," and V as "l." In Algonquin-Cree, VII, read left to right, is *Ti-ani,* equated with *Tipane,* or 'Lord.' So, it's an open question, all right."[13]

Fig. 14.12. The older profile, imperial Roman hairstyle, and lack of any inscription save the Alexander Helios sign uniquely represented in bold letters suggest this portrait stone from the cave may depict Alexander Helios himself. Photograph by Beverley Moseley

Fig. 14.13. An unusual profile from the cave collection, it is identified only by the Yahweh sign and the Alexander Helios glyph at the top.

Fig. 14.14. Accompanied by a brief Numidian inscription and the Alexander Helios sign, this soldier faces the Egyptian hieroglyph meaning "heaven."

Boutet was not the first professional investigator to detect an Iberian presence among the southern Illinois pieces. Historian Warren Cook concluded, in a letter to Russell Burrows on August 20, 1989:

> Everything I continue to learn from studying the subject confirms the probable accuracy of my deduction that the cave was a mausoleum for burying the noble dead of a Libyan-Iberian trading colony whose last Old World sovereign was the historically known King Ptolemy of Mauretania, the son of Juba II and Cleopatra Selene, daughter of Cleopatra and Marc Antony.[14]

Although the warriors portrayed on the stones are distinctly European, the non-Roman identity of at least some of them shows in their beards. Roman soldiers, and most Roman males, were usually clean-shaven until the reign of Emperor Hadrian, in the second century (fig. 14.15). So, too, the profiled heads adorned with ceremonial Egyptian regalia are not those of Egyptians, but of Mauretanians in the act of worshiping the Nile Valley deities, especially Isis. As Covey points out, "[T]he religion of Carthage was a punicized version of Egyptian sun worship (Ba'al Hammon=Lord Amon). Carthaginians evidently carried on their religious ceremonies in ritual Egyptian."[15]

The Mauretanians also used Egyptian ritual language, even though

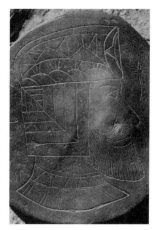

Fig. 14.15. Some warriors portrayed on the cave stones are bearded.

they evidenced some Phoenician heritage, as did King Juba II, who was directly descended from the Carthaginian occupiers of his native Numidia. Because the Mauri and Spaniards were both European, with similar defining facial structure, their portraits may be best distinguished from one another by the written languages included on them, Phoenician for the former, Iberian and Ogham for the latter.

The Phoenicians are apparent for their Semitic cast of facial features, together with their hairstyle (a short ponytail) and conical turbans (see figs. 14.16 and 14.17). And Jewish portraits are similarly recognizable for their distinctive yarmulkes, long hair, accompanying Hebrew text, and representations of holy objects such as Torah scrolls, Stars of David, and menorahs (see figs. 14.20–14.22).

Several inscribed stones of special interest feature evidence a forger would be unlikely to know. On one, a bearded man above four paleo-Hebrew letters and a Star of David wears his hair styled in seven long braids (see fig. 14.20); this is the same sign of high initiatory rank still worn by certain followers of Orthodox Judaism.

New York researcher Zena Halpern pointed out that some of the stones depict menorahs standing on triangles (see fig. 14.21). She learned while visiting an archaeological site in Israel that triangular bases were not used after the first century, an apparently insignificant bit of information only recently discovered that fits perfectly the dating of the artifacts from the first century.[16]

A particularly convincing object taken from the Olney-area site is a black prayer stone crafted to fit the hand so that the holder's thumb can caress the incised words as they are spoken (see fig. 14.23). At the top is a Star of David, beneath which, in identifiable paleo-Hebrew, is the word *Juda*.

Fig. 14.16. The Phoenician identity of this priest of Isis is suggested by his ponytail and the vertical Punic inscription at left. The horizontal inscription at bottom is Numidian, the native language of King Juba II.

A Rooster Speaks   189

Fig. 14.17. These two stone portraits depict Phoenician men, as evidenced by their ponytails and, on the stone at left, the man's Near Eastern turban with a Punic inscription.

Fig. 14.18. Full-face portraiture was a hallmark of Carthaginian artists. Here, a remarkably lifelike example from the cave is among its finest specimens. Photograph by Beverley Moseley

Fig. 14.19. The Carthaginian identity of this face is underscored by the Punic inscription at its left.

Fig. 14.20. The seven long locks in the hair of this figure from the southern Illinois site identify him as a Hebrew scholar. Note the representation of a Torah scroll above his head.

Fig. 14.21. Yarmulke, Star of David, and menorah combine on this stone from the cave to affirm the profile's Hebrew identity.

Fig. 14.22. A variation on the seven locks—the hair drawn back in a braid—as shown in this profile from the cave collection indicates this man is an advanced student of Hebrew, though not yet a scholar.

Fig. 14.23. A Hebrew prayer stone from Burrows Cave

Sometimes Jewish and Christian imagery merges on a single stone. During the mid-first century, when the Mauretanian expedition sailed to North America, Roman authorities considered Christianity a largely Jewish affair because most of its leaders, such as Saint Paul, were Jews. In fact, early Christianity attracted almost equal numbers of Jews and Gentiles, so their shared presence on one of the cave's stones corresponds to the times in which they were incised.

The Christian stones, while few, are among the most potentially dramatic of the entire collection. The death of Christ is a common artistic theme, but its depiction on the cave's stones differs from other portrayals in one important detail. Throughout medieval, Renaissance, classical, and even modern sacred art representing his execution, Jesus is invariably shown with nails driven through the palms of his hands. In all of the cave illustrations of the Crucifixion, however, the nails are through his wrists (see fig. 14.24). Only toward the close of the twentieth century did Christian scholars generally concur that had a crucified person been nailed through the palms, the body's weight would not have allowed him to maintain his position on the cross. It is conceivable, therefore, that at least a few of the Christians who sought refuge in the religious freedom of Cleopatra Selene's contemporary Mauretania could have seen and heard Jesus preach. They may even have seen the Crucifixion, which took place about ten years before they sailed for

Fig. 14.24. In this apparent representation of the death and resurrection of Christ, note the hand with a nail mark in the wrist and, over the radiant tomb, the three "suns" signifying, perhaps, the three days Jesus lay in the sepulcher. The inscription appears to be paleo-Hebrew.

Fig. 14.25. Could this portrait stone from the southern Illinois cave be a likeness of Jesus drawn from life?

America. Perhaps Jesus' nailed wrists depicted on the cave's stones were incised by an eyewitness to his death.

Even more provocative are five or six illustrated stones that seem to portray Christ himself, sometimes shown strangely hooded. The most unmistakable example shows the right profile of a long-haired, bearded man with aquiline facial features (fig. 14.25). Behind his head is a cuneiform glyph—IH/—that language scholar David Allen Deal, in his exhaustive study of this symbol, has convincingly translated as the Hebrew word *Yahw* or *Yahweh,* better known as Jehovah, literally "I am."[17] Below the Yahweh glyph is a cross (not a crucifix), flared at the ends. Below this is the ΛII meaning, perhaps, Alexander Helios. At the very bottom are a dozen vertical strokes, which could signify the twelve apostles. It seems likely that the man portrayed is meant to be Jesus Christ himself. If not, perhaps he is one of his apostles, or merely a follower. If the profiled man really is Jesus, might this illustration have been made by an artist who was his contemporary and who, perhaps, saw Jesus before his death?

According to Wayne May, Russell Burrows "purposely withheld some inscribed stones from sale because of the imagery they featured; namely, identifiably Christian scenes, mostly Old Testament. He was uncomfortable with these items because he feared critics would use such obvious themes to further debunk his discovery; they might accuse him of having copied familiar images onto the alleged artifacts. While he believed such imagery found on the stones to be genuine, he knew it would seem incredible to others. What concerned him was, as he put it, the 'Jesus stones.'"[18]

Black shipmates are also presented in enough realistic detail on the cave stones to sometimes define their origins and duties. Many are depicted with the same kind of facial scarification still found adorning the faces of males in Ghana and Senegal (fig. 14.26). At least one

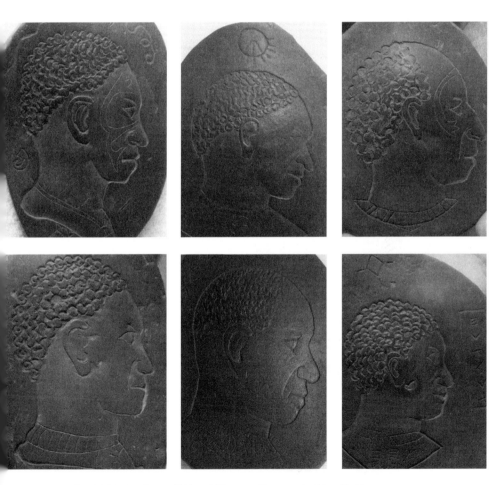

Fig. 14.26. A gallery of West African profiles may be identified by apparent facial scarification and straight neck-to-forehead haircuts.

well-preserved stone portrait shows a West African wearing a sailor's cap, with a ship included in the background (fig. 14.27).

Interestingly, on the dozen or so known stones depicting those who may be West Africans, the inscriptions are Numidian, Mauri, Phoenician, or Celtiberian (fig. 14.28), because to date there is no evi-

Fig. 14.27. This image may portray a sailor from West Africa.

Fig. 14.28. These portraits of men who seem to be West Africans are accompanied by Numidian, Phoenician, and Celtiberian inscriptions.

dence that West Africans possessed a written language of their own. "The languages on the stones and much else fit this scenario," writes Covey. "If that is not what happened, we still have to account for Mauretanian languages there and Carthaginized Egyptian religion, etc., otherwise pretty inexplicable."[19]

A small number of illustrated stones feature mythical creatures and strangely attired humans, often in bizarre circumstances. These fantastic scenes of occasional nudity and seeming ritual are in sharp contrast to the static portrayals of serious, fully clothed soldiers, sailors, or holy men on the rest of the stones. The so-called Rooster Stone has already been mentioned as a possible representation of the Gnostic supreme god, Abraxas. Other incised depictions show a man with long pointy ears, forked beard, and the legs of a goat, holding a snake (fig. 14.29).

In an article published four years before the cave was discovered, *Fate* magazine author Frank Volkmann wrote, "One [Gnostic] sect, the Ophites, became infamous for worshiping Christ in the form of a serpent." He goes on to observe that "the ideas that liberated the Gnostics from the world and its maker also liberated them from conventional

Fig. 14.29. Apparent Gnostic imagery appears on several stones from the Cave. Photographs by Beverley Moseley.

morality; their behavior was either extremely ascetic or extremely licentious. Certain sects held their services in the nude, and others may have practiced sacred prostitution."[20]

Relevant to Volkmann's comments, one stone depicts a woman, seemingly nude except for a roosterlike mask over her head, picking something from the hand of the same Pan-like figure seen holding a snake (fig. 14.30). Perhaps the figure is a Gnostic initiate receiving esoteric knowledge from the wild spirit of instinctual revelation. Her Abraxas mask and the Pan-like figure's stance and expression of seeming confidentiality may suggest the secrecy so central to Gnostic ritual.

Fig. 14.30. What seems to be a Gnostic spirit of secret wisdom holds in his left hand some sort of measuring device that may signify a "rule." Photo by Beverley Moseley

It's possible that virtually the entire collection of seven-thousand-plus cave stones might be Gnostic. Volkmann points out that "stone medallions, called Gnostic gems, have been found by the thousands in Egypt and other Mediterranean countries. No scientific or systematic study of these talismans has ever been made, so their meaning remains conjectural."[21] Precisely the same may be said for the Illinois cave stones, which are likewise mostly oval shaped and covered with enigmatic images. Their incised portraits may not only represent the men and women who fled Africa across the sea just ahead of the Roman legions, but could also depict Gnostic symbolism, which itself might be indicative of a religious conversion that took place among the spiritually diverse refugees.

Other stones are emblazoned with images of American Indians (figs. 14.31 and 14.32), suggesting that relations between the foreigners and local native peoples were common. Covey writes that "Nearly every Indian tribe in what became the U.S. was constantly at war with other tribes into the eighteenth century."[22] Remarkably, however, the tribes encountered by early American pioneers in what is now Illinois were regarded as non-belligerent indigenes. These were the Illini, Kaskasias, Michigamier, Mascotins, Cahokias, Peorias, and Taumar-waus—Algonquins all, but known collectively as the Illinois by eighteenth-century French explorers and fur traders. If the ancestors of the Illinois were just as peaceful, this may be part of the reason the Mauretanians chose southern Illinois as a final place to settle and build their royal tombs. Such information might even have been available to them

Fig. 14.31. This portrait of a Native American is accompanied by Celtiberian Ogham and Phoenician script, including, at top, the Hebrew sign for Yahweh.

Fig. 14.32. Native Americans portrayed on the Burrows Cave stones may be ancestors of Yuchi Indians, who preserved a folk memory of the cave into the twentieth century.

before setting out from Africa, guided as they were by the classified records of Carthaginian sailors, who, centuries earlier, explored the river systems of North America and used them to bring back rare trade goods.

Another native people who lived harmoniously with the ancestral Illinois were the Yuchi. In the seventeenth century, they were driven out by the more aggressive Iroquois, and resettled in the vicinity of Columbus, Georgia. Two hundred years later they were again forced to leave, like the Cherokee, by way of the notorious Trail of Tears, ending up on an Oklahoma reservation. Absorbed by the larger numbers of Cherokee, the Yuchi lost their tribal identity, although a dwindling membership still exists. Their last elected chief, Samuel W. Brown Jr., preserved a folk memory of his people's ancient history, which told how bearded, white-skinned foreigners long ago arrived in southern Illinois. There, he said, they dug tombs for their most honored dead, interring with them much gold and a library.

The chief recounted this oral tradition to archaeologist Joseph Mahan twenty-six years before Russell Burrows found the cave. Brown went on to say that Yuchi leaders ruled southern Illinois peacefully as *zopathla,* or "sun kings." He said that the Yuchi were originally known as the Yuqeeha, "sky earth people," and the Zoyaha, from *yuchi zoya,* or "sun-filled." These associations recall a historical figure who may have been Alexander Helios, as well as the Egyptian solar deities of his people and the numerous visual references to the sun found throughout the southern Illinois collection.

15
Lost Coins and Buried Treasure

*The first duty of a great historian is to be an artist.
Uninterpreted truth is as useless as buried gold. And
art is the great interpreter.*
LYTTON STRACHEY, THE SCIENCE OF HISTORY

Ever since the first American settlers arrived nearly two hundred years ago, southern Illinois has been popularly known as Egypt Land. Illinois historian John Allen writes, in his thorough examination of the origins of regional myths, that the land was so called because of its supposed resemblance to the Nile Valley, or due to the Egyptian-sounding names used by the Illini themselves to describe the land.[1] If this is so, how did the Indians first acquire these names? In truth, there is little if any resemblance between southern Illinois and the Nile Valley, with which, in all events, the largely uneducated nineteenth-century pioneers would have been almost if not absolutely unfamiliar.

In an early song they sang while entering and crossing the newly opened territories of southern Illinois, we can hear intriguing references, however coincidental, to a pre-columbian people who might have preceded them. One verse ran, "The Queen of Sheba came here with King Solomon of old, with a donkey-load of diamonds, pomegranates, and fine gold. And when she saw this lovely land, her heart was filled with joy. Said she, 'I sure would like to be a queen in Illinois.'" The song was one from a collection of pioneer ballads published in the 1950s by Illinois poet Carl Sandburg.[2]

What could have induced the anonymous composer of this ditty in the early 1800s to use such imagery as queens and kings with donkey-loads of diamonds and gold, unless there were already Indian traditions of biblical-like royalty in the area? Curiously, its verse does mention that an ancient world king and queen came here. In fact, reports of southern Illinois farmers plowing up Old World weapons and coins have persisted for generations. Perhaps the song was occasioned by early pioneers who sometimes stumbled over strange objects in the virgin territories.

Chief among these objects, interestingly enough, were Roman-era coins found by ordinary people in the course of everyday activities. Most such finds are dismissed out of hand by skeptical professionals as faked or mistaken evidence. But, as the encyclopedist Richard Corliss observes, "Almost from the time of the first American settlers, people have been discovering old coins in unlikely places. Roman coins, especially, have turned up in farmers' fields, on beaches, and elsewhere across the country. It seems that the Romans and other pre-columbian peoples either strayed far beyond the Gates of Hercules [the Straits of Gibraltar], or a lot of numismatists had holes in their pockets."[3]

One of the better-documented discoveries of its kind was described in an 1882 issue of *Scientific American* magazine. It told of a farmer who picked up a bronze coin at his property in Cass County, Illinois, about 135 miles northwest of Burrows Cave. It was examined by a St. Louis professor of classical civilization, F. R. Hilder, at the University of Missouri, who told a reporter from the *Kansas City Review*:

> Upon examination, I identified it as a coin of Antiochus IV, surnamed Epiphanes, one of the kings of Syria, of the family of the Seleucidae, who reigned from 175 to 164 B.C., and who is mentioned in the Bible

(first book of Maccabees, chapter 1, verse 10) as a cruel persecutor of the Jews. The coin bears on one side a finely executed head of the king, and on the obverse a sitting figure of Jupiter, bearing in his extended right hand a small figure of Victory, and in his left hand a wand or scepter, with an inscription in ancient Greek characters—*Basileos Antiochou Nikephorou;* the translation of which is: "King Antiochus, Epiphanes [Illustrious], the Victorious." When found, it was very much blackened and corroded from long exposure, but when cleaned, it appeared in a fine state of preservation and but little worn.[4]

A more recently found specimen was discovered shortly before World War II by a farmer working his field near Independence, Missouri. He saved it as a curiosity and never thought of showing it to a numismatist. After he passed away, his daughter allowed its photograph to be published in *Ancient American* magazine (vol. 6, no. 37).[5] Covey commented on its obvious authenticity, its apparent representation of Alexander the Great, and the inexplicable inclusion of the Greek letter *H*. Could it have stood for Alexander Helios? The nineteenth-century Antiochus IV artifact displays themes featured in the story of Mauretanians in prehistoric America, and was probably carried by a Syrian or Jew in company with other immigrants.

Artifacts such as the Illinois and Missouri coins are part of limited but revealing physical evidence for the Mauretanians' long-vanished sun kingdom, the ruins of which may still partially stand in a series of stone walls stretching from near the Mississippi River to the Ohio River. They form a broken, staggered line south of Carbondale, Marion, and Harrisburg, the three cities of the region separating the southern tip of Illinois. The locations of nine of the walls have been positively identified, and doubtless more were destroyed to make way for modern development or are yet to be uncovered in the thickly overgrown areas of the Shawnee National Forest. One of the more recent examples was found in 1970, three miles east of Cobden in Giant City State Park near Makanda (see fig. 15.1). The Lewis Wall is virtually identical to its counterparts. As Allen points out, "An inspection of this structure gives a clear idea of the general plan followed for all of them."[6]

The Lewis Wall bisects the top of a steep cliff, running on a linear east–west axis for 285 feet. Six feet tall at its highest point, with an average thickness of five feet, the structure is a dry-stone rampart containing an estimated forty thousand stones, all of them apparently conveyed by hand up the sheer incline from the dry streambed two hundred feet below. Stone cairns, or ceremonial rock piles, and pits appear at the rear entrance. The structure was raised ingeniously by fitting together mostly flat stones chosen for moderate size and a rough although uniform fit, the same technique used in building the other walls. "They were not insignificant structures," Allen insists, "and the amount of manual labor required was great. The many thousands of trips necessary to be made from the brook bed to the top of the bluff, often two hundred feet or more above the creek level, represent a stupendous effort for primitive people—the more so when it is considered that some of these walls were six hundred feet long."[7]

Archaeologists are unable to associate the Lewis Wall with the Indians, who never engaged in large-scale stonework. Nor do any Native American tribes claim it was the labor of their ancestors. Thanks to the availability of organic material embedded in the structure (material not present on the black stones of Burrows Cave), the wall has been radiocarbon-dated to the mid–first century, the same time period in which the Mauretanians' transatlantic expedition arrived in North America. That this monumental partition has stood fundamentally intact for the last twenty centuries in this major earthquake zone is testimony to its skillful construction. In 1812, the region suffered the most powerful earthquake in U.S. history, when the New Madrid Fault generated enough seismic violence to change the course of the Mississippi River and ring church bells in faraway Virginia. Sadly, farmers have pillaged its stones for building materials since the early nineteenth century, rendering a true conception of its original condition difficult to determine. Other examples, like the wall near Stonefort (a town obviously named after the structure), were more than twice as long, stretching more than 650 feet.

Regional author Judy Magee suggests that the structures may have been originally much larger:

Lost Coins and Buried Treasure 203

Fig. 15.1. Makanda's Lewis Wall may be one of the military works built by the Mauretanians for their sun kingdom.

These walls, extending east and west from the gateway, were really one wall almost a quarter of a mile in length. Some believed that this wall and the one around Old Stone Fort on the same mountain were originally eight feet high, while others believed they were as much as ten or even twelve feet high, especially near the gateways and at the ends.[8]

Writing of this particular wall, investigator Loren Coleman marveled at its formation of an accurate ellipse with axes of 450 and 190 feet. "It is not easy to see how it could have been so accurately laid out," he writes, "if the area were as heavily forested at the time as it is now."[9] Clearly, the organized labor and surveying techniques necessary to construct so many massive ramparts belonged to few if any Native American cultures in the area, but certainly typify the engineering abilities of Roman-era builders.

The Lewis Wall may once have belonged to a network of dozens of similar fortifications that defined the southern boundary of the ancient sun kingdom mentioned by Chief Brown of the Yuchi and, if so, represented the last realm of Alexander Helios. The location of Burrows

Cave is north of the line of stone walls, which all stand on top of fingerlike high bluffs thrusting southward, from which they are unassailable. Sheer cliffs fall away on either side, but each structure is easily reached from the north over gently sloping ground.

Allen points out that "a man could scale those cliffs only by careful and strenuous effort. They thus have many of the characteristics that would make them into desirable forts."[10] Their positions appear to have been not only chosen for specific environmental qualities, but also somewhat modified to meet certain military requirements. The erection of this first-century Maginot Line might well have been aimed at keeping away hostile parties. Any attackers would naturally want to strike up the Mississippi and Ohio, but their passing would have been effectively blocked at these easily defended rivers. These might be seen as very Roman modes of defense for the Roman-trained Mauretanians. But what of the stone cairns at the wall entrances? Might these have been constructed for the Celtiberians who made the voyage?

Beyond these wall fragments, the rest of the sun kingdom must have disintegrated long ago. Southern Illinois's abundance of forest timber but scarcity of building-stone quarries could well have compelled the Mauretanians to construct their homes, temples, and palaces of wood. Then, as now, builders made their structures from the local material at hand. In the event of a lost war or lethal epidemic, either of which could have forced abandonment of the kingdom and dispersal of the inhabitants, the buildings—reduced to ashes or left to the natural processes of neglect and decay for tens of centuries—simply would have dissolved without a trace.

Fig. 15.2. A replica of a ritual libation cup bearing the circle cross from the ceremonial center of Cahokia in southwestern Illinois

What may be an important piece of evidence that did not perish is a symbol still encountered, however infrequently, among some of Illinois's prehistoric sites. This is the circle cross—an O with a cross inside. What appears to be a ritual cup excavated at the great ceremonial city of Cahokia, on the east side of the Mississippi River from St. Louis, is emblazoned with the circle cross, an obvious solar image found around the world. Hence, archaeologists interpret the sign as indicative of the sun or a sun deity. Cahokia means "city of the sun," and although its greatest florescence occurred from around 900 A.D. to 1200 A.D., occupation at the site extended many centuries before then.

The circle cross is also found in another particularly sacred location, again near the Mississippi, but hidden by dense overgrowth at the base of an immense bluff, just outside the little southern Illinois town of Gorham (fig. 15.3). It is at the center of a series of petroglyphs that seem to converge on the symbol. Depictions of men being transformed into birds identify the site as perhaps a place of initiation for students of shamanism.[11]

Other circle-cross petroglyphs are found farther up the Illinois side of the Mississippi: in Pope County, on the west side of Big Hill, about four miles north of the town of Grand Tower; on the north side of the hill about a quarter mile east of Fountain Bluff; and in Jackson County, at Turkey Track Rock. Elsewhere in Illinois the glyph is less frequently seen, but it profusely adorns many artifacts retrieved from Burrows Cave. The circle cross appears as jewelry (mostly necklaces and earrings) worn by a majority of the men and women portrayed on the stones (see fig. 15.4). It is sometimes shown in an unmistakable solar context, and

Fig. 15.3. The circle-cross petroglyphs outside Gorham, Illinois

seems only somewhat less prominent than ΛII, with which it is occasionally associated. The ubiquity of the circle cross throughout the cave collections defines the symbol as a possible link between the Mauretanians and regional prehistoric tribal peoples who adopted it for their own religious use.

Allen describes other rock art at Big Hill as "geometric designs that may be attempts to make a map."[12] Pre-columbian maps in this part of North America are rare to nonexistent. Yet, the suggestion of just such a design in southern Illinois complements the several so-called map stones Russell Burrows found in the subterranean site (fig. 15.5). They are all rather primitively executed representations of what appear to be the Mississippi River valley, with tributaries stretching north into Wisconsin, east to the cave's general location, and beyond to the Appalachian Mountains. Dots between destinations seem to signify travel times in twenty-four-hour periods.

Fig. 15.4. A circle cross flanked by the elements of the Alexander Helios symbol, worn by a man who appears to be a Mauretanian soldier. Photograph by Beverley Moseley

Fig. 15.5. This map stone taken from the southern Illinois cave indicates its location on the Embarras River. Also defined is the location of Cave-In-Rock. (See upper right corner.)

The maps may indicate different cave sites. One on the north side of the Ohio River near the bottom of southern Illinois corresponds to Cave-In-Rock, a large cavern protected by the Illinois State Park System and today a popular natural attraction. Josiah Priest, a careful scholar and well-published antiquarian, whose research is still quoted by historians, visited the site in 1833. "On the Ohio, twenty miles below the mouth of the Wabash," he reported, "is a cavern, in which are found many hieroglyphics and representations of such as would induce the belief that their authors were, indeed, comparatively refined and civilized."[13] Its walls were covered with dozens of pictoglyphs, or painted illustrations. Some depicted "the sun in different stages of rise and declension," recalling the solar theme found throughout the Burrows Cave artifacts.

There was the incised drawing of "a snake biting its tail representing an orb or circle." Although not found anywhere else in the Americas, this was the *ouroboros,* an Old World symbol for the process of infinity, in which life perpetually and simultaneously devours and re-creates itself. It was the emblematic serpent of ancient Egypt and Greece. *Ouroboros* is Greek for the Egyptian *kneph,* envisioned as a snake encircling the world. It was also used by early Gnostic Christians to express the unity of all material and spiritual things, which are never really annihilated, but instead feed on each other's energies in a continuous cycle of death and renewal. Such concepts were deemed heretical by Church fathers, who anathematized any Christians who espoused them. The appearance at Cave-In-Rock of this unique symbol, in view of its cultural and spiritual associations with elements featured on the Burrows Cave artifacts (some of which are emblazoned with the ouroboros), forms a validating connection between the two sites, while providing credible evidence for the arrival of visitors in southern Illinois from across the Atlantic.

Among the rock art in Cave-In-Rock was the illustration of a crocodile—also the family emblem of the Ptolemies, to whom Cleopatra Selene and Alexander Helios belonged. "Besides these," Priest reported,

> there were several fine representations of men and women, not naked, but clothed; not as the Indians, but much in the costume of Greece

and Rome. The dress of these figures consisted of a *carbasus,* or rich cloak; a *sabucala,* or waistcoat or shirt; a *supparum,* or breeches open at the knees; *solea,* or sandals tied across the toes and heels; the head embraced by a *bandeau* crowned with flowers. The dress of the females carved in this cave have a Grecian cast, the hair encircled by the crown, and confined by a bodkin. The remaining part of this costume was Roman. The garments, called *stolla,* or perhaps the *toga pura,* flounced from the shoulders to the ground. An *indusium* appeared underneath. The *indusium* was confined under the breast by a *zone* or *cestue*. Sandals were worn in the manner of those of the men.[14]

The details of Priest's observations were verified by a famous explorer, William Pidgeon, who visited Cave-In-Rock fifteen years later.[15] Crawling into its deepest recesses, he found pictoglyphs not mentioned by his predecessor, describing them as human figures dressed as ancient Egyptians. The descriptions of these two eyewitnesses perfectly match human portrayals covering the black stones removed by the truckload from the Olney-area cave, one hundred miles to the north. It seems possible that the depictions of men and women that Priest and Pidgeon saw inside Cave-In-Rock were made by the same people responsible for the Burrows Cave artifacts. They might even have been the Mauretanians, with their mix of Ptolemaic, Greek, Egyptian, and early Christian influences.

But Cave-In-Rock had always been a haven for violent criminals, even into the early 1930s, when its walls were expunged of all rock art by local authorities in their efforts to transform it into a wholesome state park. "In WPA days," explains author Magee, "the cave at Cave-In-Rock was cleaned. Many skeletons and numerous artifacts were found in this midden. This cave, as well as others in the area, was a home for countless Indians for many hundreds or thousands of years before white men ever saw the rocks and rills of southern Illinois."[16]

It seems that more than Indians might have used Cave-In-Rock in ancient times. But any chance of following Pidgeon's crawl into the remotest interior that he saw decorated with Egyptian-like illustrations was lost forever in the 1950s, when a large portion of the ceiling collapsed, permanently blocking the far end of the cave. In any case, he and Priest recorded their personal encounters with the pictoglyphs that corresponded so closely to the culturally mixed imagery of the incised black tablets taken from the Olney-area cave.

Interestingly, the Burrows Cave map stones indicate another cave site, this one more than two hundred miles away. Strange though it may be, an unusual cave does indeed exist at the location specified by the artifact. Known as Burton Cave, it is found near the Illinois banks of the Mississippi River in the town of Quincy. Local Indians regarded the site with religious awe as "a spirit dwelling," but it was not investigated until 1885, when amateur explorers found a monumental altar stone deep inside the opening. On it lay an Egyptian-like sarcophagus with one candlestick each near the head and foot.[17] Predictably, these items disappeared when word spread of their discovery, and if Burton Cave contained anything else, it was looted. Nevertheless, this account, preceding the discovery of the Tombs of the Embarras by almost one hundred years, reflects favorably on the credibility of both sites by underscoring the map stones' veracity.

Their credibility was further confirmed in 1995 by William Kreisle, of Ohio's Midwest Epigraphic Society, when he studied two specimens that seemed geographically correct, save for their depiction of the Mississippi River. Consulting professional hydrologists, geologists, and research scientists at the Corps of Engineers in New Orleans and the Waterways Experimental Station in Vicksburg, he learned that the lower Mississippi River made an abrupt turn to the east nineteen hundred to eleven hundred years ago. Its course as portrayed on the map stones exactly conforms to the configuration of the Mississippi River as it appeared in the first century, before it changed direction, and just when the Mauretanians commenced their mass migration from Africa (see fig. 15.6).

Kreisle concluded:

> This rather esoteric knowledge regarding the eastward turn of the lower Mississippi from its original southeasterly direction, as depicted on the map stones, would hardly be widely disbursed today, thus minimizing possibilities of anyone manufacturing bogus artifacts. If

anything, the appearance of clearly recognizable river systems in the Midwest generally and Illinois specifically, as the stones indicate, go a long way toward establishing the discoveries in the cave as genuine."[18]

After the coins, the walls, the circle crosses, and the map stones, a final category of physical evidence includes the gold objects, all of them featuring symbols and images that suggest their ancient authenticity. Although most of the gold pieces seem to be in the form of currency, on closer examination they resemble commemorative medallions more than minted coins. A particularly revealing example is emblazoned with the head of an elephant. As mentioned, Cleopatra the Great had made medallions that depicted an elephant head, the symbol she designated for her reign. When presented full-bodied, the elephant was the national emblem of Mauretania under Juba II and his son King Ptolemy. Representations of elephants also appear on both coinage and tablets from Burrows Cave. Appropriately, the animal portrayed on the southern Illinois artifacts is not the large-eared sub-Saharan elephant, but rather the short-eared pygmy elephant that roamed northwest Africa until its extinction about fifteen hundred years ago. Similar to the elephant, the aforementioned crocodile images seen by Josiah Priest at Cave-In-Rock and on the stones removed from the Olney-area cave may represent the badge of the royal house of the Ptolemies, whose last survivors ruled Mauretania and escaped to prehistoric America.

Fig. 15.6. One of numerous map stones removed from Burrows Cave shows the correct configuration of the Mississippi River system as it existed two thousand years ago.

The most common icon found on these objects, however, is a triangle surmounted by a horizontal line with a circle on top, resembling a human figure wearing a skirt (see fig. 15.7). This is identical to the symbol of the Punic moon goddess, Tanit, reproduced in great numbers as votive pieces and on stelae, upright stone slabs, recovered during archaeological expeditions around the Mediterranean, particularly along the shores of the Near East and North Africa.

The Greeks knew Tanit as Astroarche, the queen of the stars, and even the Romans celebrated the deity of their enemies on the Ides of March at the *Liberalia,* where she was honored as Libera, the goddess of Libya. On one golden coin or medallion from the cave, the Tanit symbol is encircled by fourteen dots (see fig. 15.8). A modern forger would hardly have known that this number is highly significant, representing fourteen nights—the midpoint of a lunation, or lunar month, dedicated to the moon goddess. Such esoteric yet appropriate evidence strongly supports the authenticity of the artifacts. Tanit forms another connection with members of the Mauretanian exodus to South America, where a stone sculpture of her may still be seen among the pre-Incan ruins of Silustani, in the high Andes of Bolivia (see fig. 15.8).

Fig. 15.7. A cave portrait of a possible Carthaginian priest wearing an earring emblazoned with the image of Tanit. The center of the priest's pectoral is adorned with a circle cross. Photograph by Beverley Moseley

Other telling images that are typically Carthaginian (and Celtic) and that appear on many gold medallions from the Cave are those of horses' heads and fern trees. This connection provides another piece of the puzzle that is the arrival of the Mauretanians in America nearly fifteen centuries before Christopher Columbus was born.

But if they indeed arrived here, what became of these ancient refugees? No artifacts or Native American folk traditions so much as hint at their fate. Storms might have blown some far off course to the shores of Florida or Mexico. At least a few might have crossed the Isthmus of Panama to settle along the northern Pacific shores of South America in Ecuador and Peru. The blacks from West Africa may have stayed behind in what is now the northeastern quadrant of Louisiana,

Fig. 15.8. At left, a gold piece from the Burrows Cave depicts the Carthaginian Tanit emblem, while the stone Tanit at right is from the Lake Titicaca site of Silustani, in Bolivia. Photograph at left by Beverley Moseley; photograph at right by William Donato

where today the Washitaw live, people who claim to be their mixed descendants. If Carthaginian precedents are anything to go by, the Mauretanians and their fellow immigrants would never have numbered more than the thirty thousand colonialists who were part of Hanno's West African expedition five hundred years earlier. And even this number may be far too large an estimate. As such, they would have had trouble holding out indefinitely against the combined forces of disease, racial assimilation, and Indian warfare.

The Yuchi and Illini were relatively peaceful, but other tribes wandering across the territory could not be depended upon for peaceful ways. Perhaps the massive stone fortifications cordoning off the southern tip of Illinois bear witness to this. Alexander Helios, who would have been seventy-six years old upon arrival here, could not have lived for very much longer in the physically demanding circumstances of southern Illinois. He may have intended to reestablish his kingdom after having made good his escape from Roman vengeance. But landing in the middle of North America would have seemed like banishment. With his death, it is likely that Helios would have been interred in his golden sarcophagus beside his twin sister, Cleopatra Selene. Finally, with the burial of Cleopatra Selene's husband, Juba II, the Cave truly would have become the Tombs of the Embarras.

Within the space of probably only a few generations, the Mauretanians would likely have vanished through the natural processes that confronted a numerically insignificant population in the vastness of prehistoric North America. But the cave itself, it seems, continued to be

Fig. 15.9. A map of southern Illinois showing what may have been leading features of the Mauretanian sun kingdom

occupied and revered as a sacred site by the Yuchi until their expulsion from Illinois in the 1700s. Perhaps they even perpetuated the carving of black stones themselves.

In the early 1990s, Russell Burrows allegedly told investigators that he removed from the Olney-area site several tablets covered with a legible script in the French language. At least one of these supposedly bore an early-eighteenth-century date. If the report is true, it makes a great deal of sense. French explorers, missionaries, and traders frequented southern Illinois at that time, and their relations with the Indians were often amicable. The Yuchi could have learned enough of the French language to inscribe it themselves, or a favored Frenchman might have been allowed to visit the cave in person. Unfortunately, no one will ever know what the alleged text related, because Burrows reportedly reduced all the French-language tablets to unrecognizable pieces with a ball-peen hammer.[19] Apparently, he was concerned that his customers might feel the modern European inscriptions would seriously detract from the credibility of the other, far more numerous and much older artifacts.

When the Yuchi left southern Illinois, the cave fell into obscurity until it was accidentally discovered by Burrows more than two hundred years later. In the interim, it seems that only the Yuchi knew of its location. And by the time their chief told his friend Dr. Joseph Mahan about it, even he had forgotten its exact whereabouts. Samuel W. Brown Jr., who died a few years later, was the last *zapathla,* the last of the sun kings. It was a title that he inherited across nearly twenty centuries from a royal refugee, Alexander Helios—Alexander of the Sun.

16

Epilogue: The Moment of Truth

The greatest of human tragedies is a theory killed by a fact.

ROMAN ADAGE

Large-scale digging equipment arrived at the Richland County site all the way from Nevada before Christmas 2001. Preliminary ground-penetrating radar surveys revealed that the gunpowder explosion of 1989 had more than imploded the whole cave entrance and filled its deep stairwell with many tons of stone rubble; much of the subterranean network had been seriously fractured. Reentering it would be a hazardous undertaking requiring patience, diligence, and caution. The weakened walls and roof might collapse if exposed to any additional stress. These unstable conditions allowed but one means of access: It would be necessary to drill a vertical shaft directly over and into the second corridor, an estimated thirty feet beneath the surface of the hill that Wayne May and his colleagues hoped concealed the buried treasures of Alexander Helios. Through this shaft, investigators would be let down into the underground corridor on ropes.

As had occurred so often before, adverse weather conditions, this time in the form of torrential rains resulting in lakes of mud, intervened to prevent the excavators from setting up their drill. Too soon, the holidays were upon them and they resolved to renew their efforts within the week after New Year's Day. On January 5, 2002, a new, forty-foot-tall drilling rig mounted on tractors arrived at the site from Iowa. Its two-man crew sank the whirling, corkscrew bit into a spot determined by ground-penetration radar as the optimum place to break through the roof of one of the underground chambers. At thirty feet down, however, water suddenly gushed up into the newly created hollow. Another attempt was made at a secondary target. Once again, the drill struck water, and Wayne May called a halt to operations.

This day's drilling had already cost his investors almost nineteen thousand dollars. Their patience and financial support were wearing

Fig. 16.1. The drilling rig moves into place over the Cave's suspected location.

thin. Over the previous two years, they had spent tens of thousands of dollars on the project, with nothing to show so far save tantalizing subsurface electronic images (which some critics argued were being misinterpreted by the investigators) and water-filled holes.

Worse, there was real concern that the explosions set off inside the original cave entrance twenty-three years earlier had ruptured an adjacent aquifer, flooding the entire subterranean complex. If so, any perishable materials that escaped looting, such as parchments, terra-cotta tablets, statues, even copper items or human remains, would have been destroyed in the deluge. Only the inscribed stones and gold might still survive. If so, then access to the cave was more hazardous than ever. It might be entered only by scuba divers descending sixty feet into an excavated pit, at the bottom of which they would have to swim through corridors and chambers illuminated only by the lights they carried with them. The holes dug in early January might have confirmed that the underground site was inundated after all. Perhaps the ancient treasures were not only under ground; now they might be under water too, greatly multiplying the difficulty and expense of their discovery, to say nothing of their retrieval.

In any case, this latest attempt, like all those preceding it, had come to nothing. Not only financial backing for future efforts, but also the indulgence of Mr. Bougham, who owned the agricultural land being torn up under the big rig's tractor treads, was running low. Sadly, May terminated all immediate plans for future operations at the Richland County site. It was indeed the location of the world's greatest archaeological cache, he still insisted. But the harsh realities of failure forced him to postpone indefinitely any effort to bring it to light.

Kristine, his sympathetic wife, observed that perhaps there are some things that are just not meant to be discovered. The declared intentions of returning the following spring with remote television cameras that could be lowered into pits to search for underground artifacts seemed anticlimactic, at best. So, too, blowing out the original entrance with another explosion to release the pent-up floodwaters, as some suggested, was less than theoretical, since Mr. Bougham would have been horrified by such a proposition.

For skeptics, the unproductive drillings justified their condemnation of the cave as a transparent hoax. Other investigators were unable to reconcile such an abundance of masterfully carved stones with the total lack of excavation results.

Throughout 2002, the dauntless Wayne May renewed his subsurface search for the original Cave entrance with the most sophisticated ground-penetration radar available. By year's end, the Future GPR 2003/4 and Rover from Oregon's Accurate Locaters, Inc. detected the faint outlines of a buried entrance. Winter conditions prevented immediate excavation, but by spring 2003, May was again readying his forces for a final attempt at discovery.

Until he achives that breakthrough, the Cave's whereabouts are as unknown today as when it was allegedly found in 1982. Unless someone else rediscovers it by accident or design, Mother Earth will keep Russell Burrows's secret even after she has covered him in his own grave, just as she has preserved for the last two thousand years the hidden resting place of Mauretania's last king and his golden treasure.

Appendix 1

A Mauretanian Time Line

51 B.C.	Juba II, prince of Numidia (today's Libya) is born.
47 B.C.	Cleopatra VII becomes Cleopatra the Great, queen of Egypt.
46 B.C.	After the fall of Numidia, a two-year-old Juba II is taken to Rome, where he appears in Julius Caesar's triumphal procession. Thereafter, the boy is brought up as a Roman soldier and scholar.
44 B.C.	While plotting with Cleopatra to set up a dual tyranny, Julius Caesar is assassinated. His friend Marc Antony shares power with Octavian, Caesar's grandnephew.
36 B.C.	Antony and Cleopatra become lovers.
35 B.C.	Royal twins Cleopatra Selene and Alexander Helios are born in Alexandria to Antony and Cleopatra VII.
32 B.C.	Antony and Cleopatra the Great formally divide the world between themselves and her children.
31 B.C.	At the battle of Actium, the forces of Antony are defeated. Octavian becomes emperor—Caesar Augustus.
30 B.C.	Cleopatra VII commits suicide; her children, taken to Rome, are raised in the imperial household.

Appendix 1: A Mauretanian Time Line

25 B.C.	Cleopatra Selene marries Juba II. They are dispatched to rule Numidia, but after two years grow weary of the task, and move on to Mauretania (today's Morocco-Algeria), where they establish a new kingdom.
20 B.C.	Augustus is so impressed with improvements in Mauretania that he names the capital after himself, Caesarea. Cleopatra Selene gives birth to her only child, Ptolemy. Sometime thereafter, her twin brother, Alexander Helios, is summoned from Rome to live in the palace at Caesarea. Having risen to cultural and economic power, Mauretania is granted semi-autonomy by the Roman senate. The country becomes a magnet for Celtiberians from Spain, Syrians, and Phoenicians, as well as a safe haven for Jews and Christians. Juba II begins to amass a fabulous treasure and organize one of the largest libraries in the ancient world.
6 A.D.	Cleopatra Selene dies at forty years of age.
14 A.D.	Augustus dies and is succeeded by Tiberius, Cleopatra Selene's former sweetheart.
23 A.D.	Juba II dies at seventy-two years of age. Ptolemy becomes king of Mauretania.
24 A.D.	Ptolemy is acclaimed by the Roman senate for skillfully suppressing serious revolt in Mauretania.
37 A.D.	Tiberius dies and is succeeded by Caligula.
39 A.D.	Envious of Ptolemy's great wealth, Caligula invites him to Lyon, France, where the king is arrested on false charges, then taken to Rome in chains. On learning of their monarch's imprisonment, his subjects riot, killing many hundreds of Romans residing in Mauretania. Ptolemy is executed in his cell.
40 A.D.	A Roman army invades Mauretania. The freedman Aedemon is in command of Mauretanian resistance.
41 A.D.	Early in the year, Caligula is assassinated. Claudius

becomes emperor. Aedemon's forces are gradually pushed southward, toward the border with West Africa. There, rather than surrender, the Mauretanians and their allies, led by the elderly Alexander Helios, the last of Cleopatra's children, hurriedly build a makeshift fleet. Into its hulls they load Juba II's treasure and library, and then set out to take their chances on the ocean, rather than face certain death at the hands of the Romans. Although some ships are lost at sea and others, dispersed by storms, come ashore in Central and South America, the majority of survivors avoid landing in the dangerous territories of Cuba, Florida, and Mexico. Instead, they sail up the Mississippi River, fighting native tribes much of the way, until the Ohio River is reached. Steering northeast, the refugee fleet enters the Embarras River, sailing into what is now southern Illinois, home of the friendly Yuchi Indians. Here, the Mauretanians excavate a tomb for their revered dead, together with a depository for Juba II's treasure and library.

42 to circa 80 A.D. Although Alexander Helios becomes the "sun king" of the refugees' new land, he dies soon after their arrival and his realm disintegrates through warfare and assimilation with various native peoples, exacerbated perhaps by diseases against which the foreigners possess no natural immunity.

1820s and 1830s While farming in southern Illinois, early settlers occasionally find Roman-era coins and other anomalous artifacts.

1956 Chief Brown, last elected leader of the Yuchi Indians, tells Georgia archaeologist Dr. Joseph Mahan the Yuchi story of how foreigners from across the sea arrived in southern Illinois, where they dug a large tomb into which they placed a golden treasure, long since lost.

1970	A stone wall, perhaps part of the Mauretanian defensive works, is found near Makanda, in southern Illinois.
1982	Russell E. Burrows, a woodworker from the southern Illinois town of Olney, accidentally finds the tomb and treasure of Juba II. Over the next seven years, he removes an estimated seven thousand artifacts, mostly stone portraits, from the site he dubs Burrows Cave.
1989	The tomb site is closed when a powerful blast collapses its entrance.
1998	Excavation of the suspected Mauretanian site, henceforth known as the Tombs of the Embarras, begins. Over the next four years, ground-penetration radar clearly reveals its chambers, but fails to detect a point of entry.
2002	An attempt to drill into the subterranean chambers is foiled by a large underground water system.

Appendix 2

Ancient Stone Maps

The following is a reprint of "In Search of Hard Evidence: Ancient Stone Maps," by Bill Kreisle, *Ancient American* 2 , no. 11 (October/November, 1995).

When my wife, Marilyn, entered my name on the membership rolls of the Midwest Epigraphic Society, I was definitely an agnostic when it came to such matters as allegedly prehistoric inscriptions in America. But after several trips to Kentucky's Red River Gorge, where I saw such inscriptions etched into sandstone walls by ancient visitors, and after considerable research into the early history of navigation and cartography, I became less skeptical. More recently, I examined some artifacts from Burrows Cave, in particular, two pocket-sized stones, each of which have carved into their surfaces maps of a river system with major tributaries. Both maps, although differing slightly, appear to depict the Mississippi River valley. After studying and comparing them to the very early history of the Mississippi River and its tributaries, I became convinced and now believe both must be 2,000 years old or older.

These "map stones" are similar in size, approximately 3.5 by 4.5 inches, weigh about six ounces, and could be easily carried in a knapsack. The two small maps cover the same territory and include the same major tributaries: The lower Ohio, Illinois, Missouri (Platte), Arkansas, White, and Yazoo, or the Big Black River. The Wabash is also shown with what appears to be Skillet Fork—a continuation of the Little Wabash in central Illinois. On the west bank of Skillet Fork, near the Cave, both maps feature a symbol resembling the letter *E* with an extra line laying on its side, which may indicate a settlement. Although the stone maps are similar in most respects, there are important differences.

Map 1 extends further north than Map 2, probably beyond Wisconsin's St. Croix River. Map 2 shows the southern tip of Lake Michigan, perhaps indicating that the creators of the maps came from the south, and were, therefore, not Norsemen. Map 1 also shows a horseshoe-like symbol near the mouth of the Missouri River, in the vicinity of an abandoned Mandan village mentioned in William Clark's account of his journeys to the Pacific Ocean. This map also shows the Missouri River (dotted) extending past the Platte River in Nebraska. Map 1 shows a slightly different course where the Mississippi flows into the Gulf. Map 2 features many other indecipherable symbols. We may venture an interpretation of some because of their shapes and relative positions on the stone.

A cavelike symbol lies in the exact location on the Ohio River where Illinois's Cave-In-Rock is found. Interestingly, this site, only seventy-five miles from Burrows Cave, was said by early settlers to contain "Egyptian-like" artifacts. A diamond-shaped symbol on the map stones is clearly identifiable. It appears to represent the Ohio River where the Saline River tow-head, once a very large island, is precisely located. The same symbol recurs at the mouth of the Wabash River, where Wabash Island is found today.* On both maps, dots along the banks may indicate the number of days it took to travel between designated landmarks. The number of dots between these known positions are roughly proportional to the river miles between them. Another symbol—an oval with a line projecting horizontally from either end— seen on both the Arkansas and Big Black Rivers, may indicate very

---

* See any large-scale map of the Ohio River or Navigation Charts, Ohio River Louisiana District— Cairo, Illinois, to Foster, Kentucky, charts 22 through 28, Louisville District Corps of Engineers, 1994.

large bayous, although the Arkansas River symbol could also indicate the White River juncture.

A satellite image of the Big Black River east of Vicksburg, Mississippi, reveals a circular configuration similar to the symbol on the maps. Except by satellite imagery or aerial photography, it would not be known. In studying the courses of the Mississippi River on the map stones and comparing them to the present-day channel, it is apparent that a noticeable change has occurred on the lower river past the mouth of the Big Black. On the stone maps the river continues in a south-by-southeast direction, staying far west of Lake Pontchartrain before running into what is probably Bayou Lafourche, the old river channel. Today, the river runs southeast past Baton Rouge, Louisiana, turning almost due east until it passes very near the southern shore of Lake Pontchartrain, then continues in a southeasterly direction into the Gulf of Mexico, almost fifty miles east of the old channel. When were the rivers running in this configuration?

To find the answer, I turned to the Corps of Engineers in New Orleans and to the Waterways Experimental Station at Vicksburg, Mississippi. The following excerpts were taken from the report *Mississippi River and Tributaries Old River Control,* part of Memorandum No. 17, "Hydraulic Design," furnished by Arthur Laurent, chief of the Hydraulics Branch, New Orleans District Corps of Engineers, Section II-2d:

> Dating the Entrenched Valley and Alluvium. The time estimates used to show the general age of valley cutting and valley filling are taken from accepted Quaternary chronology based on worldwide belts of glaciation and related phenomena.
>
> Specific dating of meander belts and other Mississippi valley features based on the rate of meander growth has been developed by Fisk, Geological Investigation of the Lower Mississippi River, 1944. Support for this dating technique has been found in more precise estimates of age based on the radiocarbon method. Three samples taken from various depths in the Atchafalaya Basin were analyzed by J. Lawrence Kulp, Lamont Laboratories, Department of Geology, Columbia University. The samples were taken at depths of 25 feet,

73 feet, and 273 feet. The age determinations agreed within the previously established geologic data.

Section III-8. Dating the Courses. Each of the Mississippi River courses in the southern part of the alluvial valley is marked by well developed meander belts which merge into a single belt extending upstream from near Vicksburg, Mississippi, to the junction of the Mississippi and Ohio Rivers near Cairo, Illinois. Studies of accretion features on aerial photography have made possible the determination of the approximate time involved in the formation of the single upstream meander belt. Through those reconstructions, the position of the river at 100-year intervals could be ascertained, and by tracing these courses downstream it was possible to establish the time when each of the meander belts in the southern part of the valley began to form. The Maringouin-Mississippi started to develop approximately 3,000 years ago; the Teche-Mississippi, 2,000 years ago; the Lafourche-Mississippi, 1,600 years ago, and the present course of the Mississippi south of Donaldsonville was first occupied approximately 800 years ago.

Section IV-3. Teche Stage. The earliest of the Mississippi River courses in this region which may be easily traced is that of the Teche-Mississippi. While in this position, the river built the Teche Bridge, which forms the western and southern boundaries of the basin. The Teche-Mississippi followed closely the western wall of the alluvial valley for much of its length . . . It is probably at this time that the Yazoo River flowed along the eastern wall of the Mississippi River (1,900 years Before Present).

Section IV-4. The Mississippi River abandoned the Teche course on the western shed of the alluvial valley in favor of a new course (Lafourche-Mississippi) adjacent to its eastern valley around B.P. 1100.

Section IV-5. The Mississippi River abandoned its Lafourche course around B.P. 800 to B.P. 600 and occupied its present eastward course past New Orleans and Lake Pontchartrain, turning southeasterly into the Gulf, as it does today.

The contents of this report were taken largely from "Geological Investigations of the Atchafalaya Basin and Problem of the Mississippi Diversion," by Harold N. Fisk, a professor at Louisiana State University and a consultant to the U.S. Army Corps of Engineers, for the Mississippi River Commission (1952). In 1944, he wrote what became a classic monograph on the geomorphology of the Lower Mississippi Valley. For the next thirty or more years, it was considered the authoritative reference on the geologic history and chronology of the area. This classic monograph was followed in 1952 by another study for the Mississippi River Commission. In these studies, Fisk developed a chronology for specific dating of meander belts and other Mississippi valley features based on the rate of meander growth. These studies would indicate that the Mississippi turned east sometime between B.P. 1900 and 1100.*

During the past thirty years, however, another generation of scientists, using new tools and new techniques, have taken a closer look at Fisk's work and found it to be lacking.† Much of this later work has been done on an interdisciplinary basis, which includes not only geologists, but also archaeologists, engineers, biologists, etc. One of the leaders of this movement has been Roger T. Saucier of the U.S. Army Waterways Experiment Station at Vicksburg. It has been authoritatively established that Fisk underestimated the time of some geologic events by as much as several thousand years.

This, of course, has given us a new appreciation for the date when the Mississippi River turned eastward and no longer followed the rough courses shown on the stone maps. In discussion of this problem, Mr. Saucier estimated the eastern course of the Lower Mississippi to be approximately two thousand years old or older.‡ Such would, in the author's opinion, date these maps to 2,500 years ago.

---

* Fisk, Harold, *Geological Investigation of the Atchafalaya Basin and the Problem of the Mississippi Diversion,* Mississippi River Commission, 1952.

† Ausin, Burns and Miller, Saucier and Snead, "Quaternary Geology of the Lower Mississippi Valley," *The Geology of North America,* vol. K-2, *Quaternary Non-Glacial Geology: Conterminous U.S.,* The Geological Society of America, 1991.

‡ Saucier, Roger T., "Current Thinking on Riverine Processes and Geological History as Related to Human Settlement in the Southeast," *Geoscience and Man,* vol. 22, May 29, 1981.

For assisting in the research of this paper, the author wishes to thank Dr. Jack Rinker, senior research scientist at the U.S. Army Topographic Laboratories, Fort Belvoir, Virginia; Arthur Laurent, chief of the Hydraulics Branch, Corps of Engineers, New Orleans District, New Orleans, Louisiana; Roger Saucier, geologist, Waterways Experimental Station, Physical Science Geotechnical Laboratory, Vicksburg, Mississippi.

Notes

CHAPTER 1   IT ALL STARTED WITH CLEOPATRA

1. Beatrice Chanler, *Cleopatra's Daughter, the Queen of Mauretania* (New York: Liveright Publishing Corp., 1934), 98.
2. Ibid., 83.
3. Ibid., 84.
4. André Bouche-Leclercq, *Cleopatra VII and the Greco-Roman World,* trans. Maxfield Honeywell (New York: Scribners, 1923), 122.
5. Michael Grant, *Cleopatra the Great* (New York: Weidenfeld and Nicolson, 1972), 109.
6. Cyclone Covey, letter to the author, 10 September 2002.
7. Beatrice Chanler, *Cleopatra's Daughter, the Queen of Mauretania,* 87.
8. Ibid., 190.
9. Ibid.
10. Jerome Carcopino, *The Life of Cleopatra* (London: Hastings House, Ltd., 1902), 194.
11. Plutarch, *Lives,* trans. Margo McNair (London: Hastings House, Ltd., 1901), 288.
12. Beatrice Chanler, *Cleopatra's Daughter, the Queen of Mauretania,* 95.
13. Ibid., 97.
14. Plutarch, *Lives,* 52.
15. Ibid.
16. Beatrice Chanler, *Cleopatra's Daughter, the Queen of Mauretania,* 110.
17. Dimitri Rostovtzeff, *The Rise and Fall of the Ptolemies,* trans. Pavel Leysitzian (New York: Grosset and Dunlap, 1960), 88.
18. Beatrice Chanler, *Cleopatra's Daughter, the Queen of Mauretania,* 129.

CHAPTER 2   MAURETANIA

1. Cyclone Covey, letter to the author, 9 March 1999.
2. Plutarch, *Lives,* trans. Margo McNair (London: Hastings House, Ltd., 1901), 108.
3. Beatrice Chanler, *Cleopatra's Daughter, the Queen of Mauretania* (New York: Liveright Publishing Corp., 1934), 121.
4. Herbert Rheingold, *Early Imperial Rome* (London: Hastings House, Ltd., 1901), 111.
5. Beatrice Chanler, *Cleopatra's Daughter, the Queen of Mauretania,* 83.
6. Reginald MacCauley, *The Horse and Horsemanship in the Ancient World* (New York: Liveright Publishing Corp., 1940).
7. Beatrice Chanler, *Cleopatra's Daughter, the Queen of Mauretania,* 132.
8. Theodor Mommsen, *The Provinces of the Roman Empire from Caesar to Diocletian,* vol. 2, trans. William P. Dickson (New York: Scribners, 1913), 262.
9. Beatrice Chanler, *Cleopatra's Daughter, the Queen of Mauretania,* 112.
10. Cyclone Covey, letter to the author, 21 August 2001.
11. Maxim Gsell, *Rome in Africa,* trans. Hubert Hollander (New York: Scribners, 1920), 313.
12. Theodor Mommsen, *The Provinces of the Roman Empire from Caesar to Diocletian,* 253.
13. A. MacCallum Scott, "The Medrassen and the Tombeau de la Chrétienne," in *Records of the Tombs,* vol. 2 of *Wonders of the Past* (New York: Wise and Co., 1923), 172, 173.
14. Ibid., 169.
15. Ibid., 172.
16. Ibid.
17. Ibid.
18. Ibid., 171.

CHAPTER 3   SPQR: FOR THE SENATE AND PEOPLE OF ROME

1. Plutarch, *Lives,* trans. Margo McNair (London: Hastings House, Ltd., 1901), 196.
2. Beatrice Chanler, *Cleopatra's Daughter, the Queen of Mauretania* (New York: Liveright Publishing Corp., 1934), 118.
3. Cyclone Covey, letter to the author, 20 April 1999.
4. Beatrice Chanler, *Cleopatra's Daughter, the Queen of Mauretania,* 79.
5. Ibid., 71.
6. Alice Curtis Desmond, *Cleopatra's Children* (New York: Dodd, Mead, 1971), 159.
7. Theodor Mommsen, *The Provinces of the Roman Empire from Caesar to Diocletian,* vol. 2, trans. William P. Dickson (New York: Scribners, 1913).

CHAPTER 4 CALIGULA: A MIND ABUSED

1. Cyclone Covey, letter to the author, 30 April 2000.
2. Hutton Webster, *Ancient History* (Chicago: D. C. Heath and Co., 1913), 274.

CHAPTER 6 ESCAPE OR DIE

1. Norman Totten, "King Juba Remembered: A Working Hypothesis," in *Across Before Columbus? Evidence for Transoceanic Contact with the Americas prior to 1492* (Edgecomb, Maine: NEARA Publications, 1998), 31.
2. Zelia Nuttall, *The Fundamental Principles of Old and New World Civilizations,* vol. 2 (Cambridge, Mass.: Peabody Museum, Harvard University, 1900), 188.
3. Norman Totten, "King Juba Remembered: A Working Hypothesis," 27.
4. Gunnar Thompson, *American Discovery: The Real Story* (Seattle: Argonauts O.T.M.I., 1992), 202.
5. Randall Wertheimer, *A Geological History of the Mississippi Valley* (New Richmond, Wis.: St. Croix Press, 1983), 4.

CHAPTER 7 DISCOVERY IN SOUTHERN ILLINOIS

1. A. MacCallum Scott, "The Medrassen and the Tombeau de la Chrétienne," in *Records of the Tombs,* vol. 2 of *Wonders of the Past* (New York: Wise and Co., 1923).
2. Russell E. Burrows, remarks made before the World Ancient History Conference in Vienna, 9 April 2001. Quoted in *Das Geheimnis der Vergangenheit* (Wien: Brestmann Verlag, 2001), 44–45.
3. Ibid.
4. Ibid.
5. Ibid.

CHAPTER 8 GOLD, ARCHAEOLOGICAL AND OTHERWISE

1. John Tiffany, "Mystery Cave Could Prove Ancient Visitors Were Here," in *The Barnes Review* 7, no. 5 (September/October 2001).
2. Horatio Rybnikar, "John Ward: Curator of the Secret," in *Ancient American* 3, no. 16 (January/February 1997).
3. Jack Ward, private papers, unpublished manuscript, 1989.
4. Ibid.
5. Ibid.
6. Ibid.
7. Harry Hubbard, "The Greatest Archaeological Discovery of the Century," in *Ancient American* 3, no. 16 (January/February 1997).
8. Ibid.
9. Ibid.

10. Ibid.
11. Ibid.
12. Ibid.
13. Ibid.
14. Ibid.
15. John Tiffany, "Mystery Cave Could Prove Ancient Visitors Were Here."

CHAPTER 9   FIND OR FRAUD OF THE CENTURY?

1. Russell Burrows and Fred Rydholm, *The Mystery Cave of Many Faces* (Marquette, Mich.: Superior Heartland, Inc., 1991).
2. Ibid.
3. Ibid.
4. Ibid.
5. Ibid.
6. Ibid.
7. Luc Buergin, *Geheimakte Archaelogie, Unterdrueckte Entdeckungen, verschollene Schaetze, bizarre Funde* (Munich: F. A. Herbig Verlagsbuchhandlung GmbH, 1998).
8. Russell Burrows and Fred Rydholm, *The Mystery Cave of Many Faces*.
9. Gunnar Thompson, *American Discovery: The Real Story* (Seattle: Argonauts O.M.T.I., 1992).
10. *Ancient American* 3, no.13 (May/June, 1996).
11. James P. Scherz, *Rock Art Pieces from Burrows Cave in Southern Illinois,* vol. 1 (Marquette, Mich.: Superior Heartland, Inc., 1992).
12. ———, " America's Pre-Columbian Monroe Doctrine: Its Origins and How It Works, as Illustrated by the Burrows Cave Saga," *Ancient American* 7, no. 9 (September/October 2001).
13. Ibid.
14. Luc Buergin, *Geheimakte Archaelogie, Unterdrueckte Entdeckungen, verschollene Schaetze, bizarre Funde.*
15. Russell Burrows and Fred Rydholm, *The Mystery Cave of Many Faces.*
16. Ibid.
17. Ibid., 54.
18. Luc Buergin, *Geheimakte Archaelogie, Unterdrueckte Entdeckungen, verschollene Schaetze, bizarre Funde.*
19. Jack Ward, private papers, unpublished manuscript, 1989.
20. Luc Buergin, *Geheimakte Archaelogie, Unterdrueckte Entdeckungen, verschollene Schaetze, bizarre Funde.*
21. Ibid.
22. Jack Ward, private papers, unpublished manuscript, 1989.

23. Wayne May, "'Mudstone Source for Burrows Cave Found," *Ancient American* 4, no. 29 (October/November 1999).
24. Luc Buergin, *Geheimakte Archaelogie, Unterdrueckte Entdeckungen, verschollene Schaetze, bizarre Funde.*
25. Ibid.
26. Ibid.

CHAPTER 10 FIRE IN THE HOLE

1. Russell Burrows and Fred Rydholm, *The Mystery Cave of Many Faces* (Marquette, Mich.: Superior Heartland, Inc., 1991).
2. "Burrows Cave: A Report," *Ancient American* 7, no. 30 (August/September 1998).
3. Ibid.
4. Ibid.
5. Ibid.
6. James P. Scherz, *Rock Art Pieces from Burrows Cave in Southern Illinois,* vol. 1 (Marquette, Mich.: Superior Heartland, Inc., 1992).
7. Russell Burrows Speaks Out on the Mystery Cave, *Ancient American* 1, no. 4 (January/February 1994).
8. Cyclone Covey, letter to the author, 23 February 1999.
9. Evan Hanson, letter to the editor, *Ancient American* 2, no. 11 (October/November 1995).
10. James P. Scherz, letter to the editor, *Ancient American* 2, no. 11 (October/November 1995).
11. George W. Lodge, "The Business of Discovery,"*Ancient American* 3, no. 16, (January/February 1997).
12. Ibid.
13. Ibid.
14. Ibid.
15. Author's private correspondence, 30 April 2001.
16. Cyclone Covey, letter to the author, 22 May 2001.
17. Ibid.

CHAPTER 11 WHERE IS THE CAVE?

1. Wayne May, "Update on Southern Illinois Site," *Ancient American* 6, no. 42 (November/December 2001).
2. Russell Burrows, interview, KABC Radio, Los Altos, Calif., 20 July 1999.
3. Russell Burrows, "Burrows Cave Is Opened!" *Ancient American* 5, no. 33 (June 2000).
4. Ibid.
5. Ibid.

6. Ibid.
7. Wayne May, "Update on Southern Illinois Site," *Ancient American* 6, no. 42 (November/December 2001).
8. Ibid.

CHAPTER 12 THE PASTURE OF FOOLS

1. Beatrice Chanler, *Cleopatra's Daughter, the Queen of Mauretania* (New York: Liveright Publishing Corp., 1934), 206.
2. Homer, *The Odyssey,* trans. Albert Handley (London: Hardyman Publishers, Ltd., 1990), 76.
3. Virgil, *The Aeneid,* book 4, trans. Violette Lloyd (London: Hammersmith, 1903), 585.
4. Ibid., (book 10, 161, 162).
5. Ibid., (book 4, 361, 362).
6. David Hatcher Childress, *Ancient Aircraft of Atlantis and India* (Stelle, Ill.: Adventures Unlimited Press, 1996), 83.
7. Alfred J. Church, *Stories of the East from Herodotus* (New York: Dodd, Mead and Company, 1900), 92.
8. Homer, *The Odyssey,* 12.
9. Lionel Casson, *Ships and Seamanship of the Ancient World* (Princeton: Princeton University Press, 1971), 58.
10. Richard Leakey, *Early Man* (London: Trever Publishers, Ltd., 1978), 162.
11. Homer, *The Odyssey,* 31.
12. Gunnar Thompson, *American Discovery: The Real Story* (Seattle: Argonauts O.T.M.I., 1992), 268.
13. Plutarch, *Lives,* trans. Margo McNair (London: Hastings House, Ltd., 1901), 144.
14. Alban Wall, "An Ancient Greek Historian's Sailing Directions to North America," in *Ancient American* 6, no. 37 (February 2001).
15. Zelia Nuttall, *The Fundamental Principles of Old and New World Civilizations,* vol. 2 (Cambridge, Mass.: Peabody Museum, Harvard University, 1913), 242.
16. Beatrice Chanler, *Cleopatra's Daughter, the Queen of Mauretania,* 197.
17. Edgerton Sykes, *Man across Atlantic* (London: Robertson Press, 1958).
18. Diodorus Siculus, *The Geographica of Diodorus Siculus,* trans. Hans Bender (Cambridge, Mass.: Harvard University Press, 1889), 239.
19. Beatrice Chanler, *Cleopatra's Daughter, the Queen of Mauretania,* 226.
20. Gunnar Thompson, *American Discovery: The Real Story,* 241.
21. Ibid., 242.
22. Bill Toma, "New Discovery of Ancient Maps Puts Phoenicians in America," in *Ancient American* 3, no. 16 (March/April 1997).
23. Ibid.

24. Gunnar Thompson, *American Discovery: The Real Story*, 174.
25. Norman Totten, "King Juba Remembered: A Working Hypothesis," *Across Before Columbus? Evidence for Transoceanic Contact with the Americas prior to 1492* (Edgecomb, Maine: NEARA Publications, 1998), 30.

CHAPTER 13 THE TESTIMONY OF THE PAST

1. Gunnar Thompson, *American Discovery: The Real Story* (Seattle: Argonauts O.T.M.I., 1992), 188.
2. Ibid., 196.
3. Ibid., 197.
4. Ibid., 126.
5. Ibid., 103.
6. Ibid., 108.
7. Ibid.
8. Norman Totten, "King Juba Remembered: A Working Hypothesis," *Across Before Columbus? Evidence for Transoceanic Contact with the Americas prior to 1492* (Edgecomb, Maine: NEARA Publications, 1998), 24.
9. Ibid., 26.
10. Ibid., 32.

CHAPTER 14 A ROOSTER SPEAKS

1. James P. Scherz, *Rock Art Pieces from Burrows Cave in Southern Illinois*, vol. 1 (Marquette, Mich.: Superior Heartland, Inc., 1992), 63.
2. Theodor Mommsen, *The Provinces of the Roman Empire from Caesar to Diocletian*, vol. 2, trans. William P. Dickson (New York: Scribners, 1913), 195.
3. Ibid.
4. Beatrice Chanler, *Cleopatra's Daughter, the Queen of Mauretania* (New York: Liveright Publishing Corp., 1934), 211.
5. Cyclone Covey, letter to the author, 21 August 2001.
6. Michel-Gérard Boutet, "Burrows Cave: A Celtiberian Cache," *Midwestern Epigraphic Journal* 11 (1997), 47–75.
7. Ibid.
8. Ibid.
9. Ibid.
10. Ibid.
11. Ibid.
12. Ibid.
13. Cyclone Covey, letter to the author, 21 August 1999.
14. Luc Buergin, *Geheimakte Archaelogie, Unterdrueckte Entdeckungen, verschollene Schaetze, bizarre Funde* (Munich: F. A. Herbig Verlagsbuchhandlung GmbH, 1998), 244.

15. Cyclone Covey, letter to the author, 21 August 1999.
16. Zena Halpern, "Jews in Pre-Columbian North America," *Ancient American* 5, no. 24 (August/September 1999).
17. David Allen Deal, "'The Mystic Symbol' De-Mystified," *Ancient American* 1, no. 5 (March/April 1994).
18. Wayne May, interview by Art Bell, *Coast to Coast with Art Bell*, 25 August 2001.
19. Cyclone Covey, letter to the author, 29 October 2001.
20. Frank Volkmann, "Secret Gospels of the Egyptian Gnostics," *Fate* 31, no. 9, (September 1978).
21. Ibid.
22. Cyclone Covey, letter to the author, 30 January 2000.

CHAPTER 15   LOST COINS AND BURIED TREASURE

1. John W. Allen, *Legends and Lore of Southern Illinois* (Carbondale: Southern Illinois University Press, 1963), 121.
2. Carl Sandburg, *Collected Works* (Three Rivers, Mich.: River Run Press, 1968), 16.
3. Richard Corliss, *Ancient Structures, Remarkable Pyramids, Forts, Towers, Stone Chambers, Cities, Complexes* (Glen Arm, Md.: The Sourcebook Project, 2001), 170.
4. Gunnar Thompson, *American Discovery: The Real Story* (Seattle: Argonauts O.T.M.I., 1992), 215.
5. Cyclone Covey, "Ancient Greek Coin Found in Missouri," *Ancient American* 6, no. 37 (February 2001).
6. John W. Allen, *Legends and Lore of Southern Illinois*, 148.
7. Ibid.
8. Judy Magee, *Cavern of Crime* (Paducah, Ky.: Riverfolk Publishing Co., 1973), 14.
9. Richard Corliss, *Ancient Structures, Remarkable Pyramids, Forts, Towers, Stone Chambers, Cities, Complexes*, 170.
10. John W. Allen, *Legends and Lore of Southern Illinois*, 149.
11. Frank Joseph, *Sacred Sites: A Guide to Mysterious Places* (St. Paul, Minn.: Llewellyn, 1992), 242.
12. John W. Allen, *Legends and Lore of Southern Illinois*.
13. Cyclone Covey, "Pre-Columbian Crucible: The Birthplace of American Civilization," *Ancient American* 5, no. 31 (February 2000).
14. Ibid.
15. Ibid.
16. Judy Magee, *Cavern of Crime*, 14.
17. Ibid.
18. Bill and Marilyn Kreisle, "In Search of Hard Evidence," *Ancient American* 2, no. 11 (October/November 1995).
19. Wayne May, interview by Art Bell, *Coast to Coast With Art Bell*, 25 August 2001.

Bibliography

Allen, John W. *Legends and Lore of Southern Illinois.* Carbondale: Southern Illinois University Press, 1963.

Barrett, Anthony. *Caligula: The Corruption of Power.* New York: Simon and Schuster, 1991.

Boutet, Michel-Gérard. "Burrows Cave: A Celtiberian Cache." In *Midwestern Epigraphic Journal* 11, 1997.

Budge, E. A. Wallis. *Egyptian Language.* New York: Dover, 1967.

Buergin, Luc. *Geheimakte Archaelogie, Unterdrueckte Entdeckungen, verschollene Schaetze, bizarre Funde.* Munich: F. A. Herbig Verlagsbuchhandlung GmbH, 1998.

Burrows, Russell, and Fred Rydholm. *The Mystery Cave of Many Faces.* Marquette, Mich.: Superior Heartland, Inc., 1991.

Casson, Lionel. *Ships and Seamanship of the Ancient World.* Princeton: Princeton University Press, 1971.

Chanler, Beatrice. *Cleopatra's Daughter, the Queen of Mauretania.* New York: Liveright Publishing Corp., 1934.

Corliss, Richard. *Ancient Structures, Remarkable Pyramids, Forts, Towers, Stone Chambers, Cities, Complexes.* Glen Arm, Md.: The Sourcebook Project, 2001.

Desmond, Alice Curtis. *Cleopatra's Children.* New York: Dodd, Mead, 1971.

Fell, Barry. *America B.C.: Ancient Settlers in a New World.* New York: Simon and Schuster, 1976.

———. *Bronze Age America.* Boston: Little, Brown and Co., 1982.

Grant, Michael. *Cleopatra the Great.* New York: Weidenfeld and Nicolson, 1972.

Hoffmann, Max. *Ptolemais von Mauretanien*. In *Paulys Real-Encyclopaedie der klassischen Altertums-Wissenschaft*. Vol 23. Frankfort-am-Main: Goerdler Verlag, 1889.

Macurdy, Grace Harriet. "Ptolemaic Queens." In *Hellenistic Queens: A Study of Woman Power in Macedonia, Seleucid Syria and Ptolemaic Egypt*. Vol. 14 of *Johns Hopkins Studies in Archaeology*. Baltimore: Johns Hopkins University Press, 1932.

Magee, Judy. *Cavern of Crime*. Paducah, Ky.: Riverfolk Publishing Co., 1973.

Mommsen, Theodor. *The Provinces of the Roman Empire from Caesar to Diocletian*. Vol. 2. Translated by William P. Dickson. New York: Scribners, 1913.

Plutarch. *Lives*. Translated by Margo McNair. London: Hastings House, Ltd., 1901.

Pritchard, James B. *The Sea Traders*. New York: Time-Life Books, 1974.

Scherz, James P. *Rock Art Pieces from Burrows Cave in Southern Illinois*. Vol. 1. Marquette, Mich.: Superior Heartland, Inc., 1992.

———. "America's Pre-Columbian Monroe Doctrine: Its Origins and How It Works, as Illustrated by the Burrows Cave Saga." In *Ancient American* (September/October 2001).

Scott, A. MacCallum. "The Medrassen and the Tombeau de la Chretienne." In *Records of the Tombs*. Vol. 2 of *Wonders of the Past*. New York: Wise and Co., 1923.

Spence, Lewis. *Myths and Legends of Babylonia and Assyria*. London: George G. Harrap and Company, 1916.

Tiffany, John. "Mystery Cave Could Prove Ancient Visitors Were Here." In *The Barnes Review* 7, no. 5. (September/October 2001).

Thompson, Gunnar. *American Discovery: The Real Story*. Seattle: Argonauts O.T.M.I., 1992.

Totten, Norman. "King Juba Remembered: A Working Hypothesis." In *Across Before Columbus? Evidence for Transoceanic Contact with the Americas prior to 1492*. Edgecomb, Maine: NEARA Publications, 1998.

Walker, Barbara. *The Woman's Encyclopedia of Myths and Secrets*. San Francisco: HarperSanFrancisco, 1983.

Webster, Hutton. *Ancient History*. Chicago: D. C. Heath and Co., 1913.